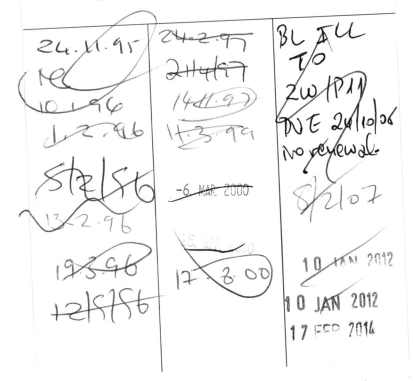

Wheat End Uses
Around the World

Edited by

Hamed Faridi

Nabisco, Inc.
East Hanover, New Jersey

Jon M. Faubion

Kansas State University
Manhattan

American Association of Cereal Chemists
St. Paul, Minnesota, USA

This book was formatted from computer files submitted to the
American Association of Cereal Chemists by the editors of the volume.
No editing or proofreading has been done by the Association.

Reference in this publication to a trademark, proprietary product,
or company name by personnel of the U.S. Department of Agriculture
or anyone else is intended for explicit description only and does not
imply approval or recommendation to the exclusion of others
that may be suitable.

Library of Congress Catalog Card Number: 95-75074
International Standard Book Number: 0-913250-87-2

Printed in the United States of America on acid-free paper

American Association of Cereal Chemists
3340 Pilot Knob Road
St. Paul, Minnesota 55121-2097, USA

Contents

Foreword

Simon S. Jackel

The interdependence between man and wheat is not limited to regional or ethnic preferences. It is truly global in scope. Wheat is being grown in all fertile areas of the world except the ultra-high temperature tropics. Orth and Shellenberger (1988) quote Percival (1921) that wheat is being harvested somewhere in the world in any given month.

Jacob (1944) interprets the sketchy evidence available that the oldest wheat originated in Abyssinia and descended into the hot river valley of the Nile. This wheat, which grew in Egypt, was hardly like the wheats that today cover the great fields of the United States, Canada, and the Ukraine. This was emmer wheat, a very early form of bread wheat. The Romans bred other wheats and established their improved wheats as the predominant grain of the Mediterranean. From Roman times on, history is written and well-documented.

The history of bread revolves upon wheat and rye—with wheat being much more prominent. Only wheat, and to a much lesser extent rye, has the protein structures required to retain gas in the leavened dough and produce a light aerated loaf of bread. Bread ruled over the ancient world. The Egyptians, who invented it, based their entire administrative system upon it. The Jews made bread the starting point of their religious and social laws. The Greeks created profound legends for the Bread Church of Eleusis. The Romans converted bread into a political factor. They ruled by it, conquered the entire world by it, and lost the world again through it. Eventually, bread took on the spiritual significance of Jesus Christ when he said "Eat! I am the bread."

Although many wheats are still in existence, and possibly cultivated somewhere, Orth and Shellenberger (1988) limit consideration of commercially relevant crops to "four species of the genus *Triticum*: *T. monococcum* (diploid), *T. turgidum* (tetraploid), *T. timopheevi* (tetraploid), and *T. aestivum* (hexaploid). Of these *T. aestivum*, which includes the common or bread wheats, and *T. turgidum*, which includes the durum or macaroni wheats, are by far the most wide grown."

These species can then be subclassified still further into hard red winter and hard red spring, according to when planted, and are used for leavened bread production and for mixing with weaker wheats to strengthen the flour. Soft red winter and common white wheat classes, along sometimes with club wheat, are mainly used for pastries, crackers, and cookies. Durum wheat is used for macaroni, spaghetti, and other pasta products.

International trade in wheat requires an understanding by all parties that the specifications have very real meanings. This may not be as easy as it sounds, mainly because importers and exporters often place their reliance not on specifications but on performance in a single, easily run functional test, such as the alveograph or mixograph.

The world production of wheat is massive in nature and continues to increase at a significant rate as the nutritional virtues of complex carbohydrates are accepted worldwide. Vast progress has been made in establishing the importance of wheat and cereal complex carbohydrates as "clean-burning" energy sources for human consumption. The growth that has taken place in wheat utilization in North America (see Chapter 1) is a good example of the increasing level of importance of wheat in Western diets.

Although wheat is consumed in many forms, including noodles, gruels, cooked cereals, and ready-to-eat cereals, bread has established itself worldwide as a major convenient and delicious food especially when made according to regional preferences for flat breads, French breads, Italian breads, Vienna bread, white and variety pan breads, pumpernickel and rye breads, Irish bread, Swedish limpa breads, Polish bread specialties, and many other ethnic varieties. With the growing availabilities of frozen doughs and prepared mixes, it is now feasible for bakeries to offer a wide variety of the worldwide ethnic breads.

Because of this improved availability, the writer proposed the following definition of a baker at the 1986 International

U.S. Wheat End Use Quality Conference in Fargo, North Dakota: "The baker converts the flour produced by the miller from wheat grown by the producer into delicious basic foods favored by the consumer and capable of being eaten with every meal, by every ethnic and age group every day of the year, regardless of economic level from the cradle to the grave."

World trade in wheat is a major factor in the political and economic interrelationships between governments. Subsidies to farmers do much to make them more competitive in the worldwide market, which is dominated by trading partners with long histories of tough trading policies. Nevertheless, although global trading is essential, the consequences in the domestic wheat markets are sometimes catastrophic. It was not so many years ago that wheat growers in the United States marched on Washington, DC, to protest the government's policies as they affected them. It was only a year or two ago that French wheat growers threatened to strike because the French government was planning to put restrictions on them as required by membership in the European Economic Community. Similarly, Canadian wheat producers protested the free import of United States wheat as specified in the North American Free Trade Agreement (NAFTA). A serious United States government decision in 1974 to meet the wheat needs of the then Soviet Union, which previously had not been a buyer on the world wheat market, resulted in cleaning out all of the reserves in the United States. These reserves have not been built up to this day.

Ancient Egyptians invented the art of leavening and baking raised bread with its lighter structure. Jacob (1944) reports that Egyptian white breads were produced as elongated loaves or as small, round loaves. Some breads were decorated with seeds. With the advent of the Industrial Revolution, baked foods of many different formulations, including cakes, crackers, and cookies, were developed and baked in many different designs of baking pans and on many different baking belts.

To this day, new developments have mainly taken the form of incorporating a variety of ingredients (sugar, shortening, fiber, bran, milk, spices, etc.), as well as incorporating other flours such as soy, potato, rye, oats, barley, etc. Vital wheat gluten has provided a major ingredient needed to improve dough strength when large amounts of nongluten containing flours and fibers are incorporated.

The use of wheat in animal feeds is nutritionally sound,

even though lysine is on the short side, but to a large extent is dependent on economic factors, since it competes with other cereals, such as soy, for space in the animal dietary.

A major other use for wheat is its separation into vital wheat gluten and wheat starch, usually by hydrating the flour to form a dough and then washing out the starch by repeated screenings and centrifugations. The availability of vital wheat gluten has turned out to be of maximum value to the baker who needs to reinforce his dough strength because stronger flour is not available or is too expensive, or because certain of his formulations, such as the reduced-calorie, high fiber breads, need to carry increased quantities of water and lack the absorption strength to do so without the addition of vital wheat gluten. It should be mentioned that wheat gluten, particularly the product which has become denatured during processing, is used by many breakfast cereal, meat, fish and poultry products, and animal feed and pet food producers to increase protein levels.

From vital wheat gluten's humble beginnings as a by-product of wheat starch production, to its role today as an ultra-important ingredient of variety, high fiber and reduced-calorie breads, there has been a vast increase in volume produced and in countries producing vital wheat gluten as the primary product of wheat separation, with wheat starch emerging as the by-product.

In the so-called emerging countries, such as those in Asia and Africa, there is a prevalent shortage of protein in the diet. The use of wheat gluten, as well as other protein sources, to enhance protein content by incorporation in local foods is being promoted actively by nutritional authorities. Wheat gluten is widely used as an extender or protein substitute in meat products.

In European countries, including the United Kingdom, Italy, France and the Scandinavian countries, which have major programs to use domestically grown low-protein wheats, vital wheat gluten is added to manufacturers formulations to strengthen the dough structure and improve quality of baked foods. The alternative to domestically produced vital wheat gluten is to import strong blending wheats from other countries, such as the United States and Canada, but importing wheats is discouraged.

In North America, there has been a wide-acceptance by consumers of variety breads. These breads tend to have weak dough structure because of the addition of non-wheat grains and flours which lack gluten. The tendency to weak dough structure, which shows up in reduced loaf volume and coarse, irregular internal crumb structure, is overcome successfully by the addition of vital wheat gluten. More than 15% of the pan bread production in the United States is in the so-called "Lite Bread" category. These are breads with at least one-third fewer calories than their regular counterparts, achieved by in-corporating as much as 22% fiber (based on total flour) and increasing water content in the baked food from the usual 34 to 38% to an enhanced level of 42 to 44%. Vital wheat gluten is successful in providing these breads with sufficient volume (lightness), interior crumb structure, and mouth feel and taste to be well-accepted by the consumer.

In wrapping up this text on the world-wide impact of wheat, we should take a look at what the future holds. There are po-litical economic factors to be considered with the European Economic Community (EC) well-launched, with the North American Free Trade Agreement (NAFTA) in the works, and the General Agreement on Tariffs and Trade (GATT) being worked out. There are environmental factors to be considered, with restrictions coming on some of the more effective pesti-cides and weed control agents (herbicides). Climatic conditions may be variable with the greenhouse effect needing to be con-sidered. Biotechnology, with its ability to introduce appropri-ate genetic material from non-wheat sources into the wheat seed nucleus itself, is just coming into its own.

Couple these factors with increasing nutritional knowledge and awareness of the basic role of wheat, and other cereals and grains in the human dietary, as evidenced by the U.S. De-partment of Agriculture Food Guide Pyramid listing the bread, cereal, rice and pasta group as the base of the pyramid, with 6–11 servings recommended daily.

The world population is aging, with its particular nutri-tional and dietary requirements; former third-world countries are now advancing and are known as emerging countries. More and more village dwellers are moving to cities, resulting in decreased rural agricultural production and increased ur-ban food requirement. The world order is changing with the breakdown of Communism. What effect will these world-wide and global events have on the world-wide role of wheat and its many strains and cultivars? An effort to project the impact of some of these factors is now in order.

The negotiations and governmental interrelationships that are required to launch uniform agreements and standards are highly delicate and complex, and yet are necessary to meet future needs of a burgeoning population. The International Wheat Council, quoted in World Grain, 1993, has estimated that world wheat consumption will grow 10% by the year 2000. An example of problems encountered in establishing broad agreements within the North American Free Trade Agreement (NAFTA) is quoted from the Independent Bakers Association Bulletin dated February 10, 1994: "U.S. and Canadian officials are stepping up negotiations for a broad agreement on durum wheat. However, in order to drum up support for NAFTA, President Clinton had agreed to launch an investigation into allegations by U.S. farmers that unfair pricing and subsidies accounted for sharp increases in imports of Canadian durum wheat. This could lead to quotas or countervailing duties against Canada."

The problems of the European Economic Community (EC) stemmed in particular from the desire to standardize all factors so that there could be Euro-Wheat and Euro-flour and Euro-Baked Foods. These efforts are now on hold because country-by-country, preferences for their own ethnic types of breads and other wheat foods overshadowed all other considerations. Nevertheless, with a European Common Market of about 350 million heads, there will be a free exchange of wheat products from one country to another, leading in due time to a wider acceptance of previously limited products and wheat specifications.

The worldwide General Agreement on Tariffs and Trade (GATT) has not yet been ratified by all 117 nations. GATT has strong support in the United States and is expected, according to the Independent Bakers Association Bulletin of February 8, 1994 "to reduce international tariffs; expand free trade; stimulate business and investment activity; and open previously hard-to-access world markets." It is obvious from these comments that it will take time for the parties to these agreements to develop working arrangements. Once this is accomplished, the needs of humankind will be better served, including producing and conserving the wheats of greatest significance and utility.

The effect of changing climatic and environmental conditions and the regulation of pesticides and herbicides are of utmost importance to the future of grain utilization. It is inevitable that environmental factors will be placed high on the agenda for global trade discussions. Each country may very well have their own views on what is and what is not acceptable to them, according to their views on the health of their citizens and the protection of the environment. Meanwhile, some comments from the United States' perspective can be made. In October 1993, the U.S. government announced a voluntary program to combat greenhouse gases which some environmentalists believe cause global warming. There is an understanding that if industry does not make voluntary progress, more stringent programs will be initiated in 1995.

According to the U.S. Environmental Protection Agency (EPA), there are eight excessive pollutants found in the "storm-water run-off" associated with the food industry. Four of the eight are background metals commonly found in the soil in very low amounts. EPA is proposing a model storm-water permit and quarterly storm-water sampling program for food industries with five or more severe pollutants in the twelve states over which it has jurisdiction. (The 38 additional states and the Virgin Islands have their own licensing authorities.)

The Clean Air Act passed in 1990, in the United States, required the 50 states to submit their plans for implementing Clean Air Regulations by November 15, 1993. The Environmental Protection Agency (EPA) by November 15, 1994 should review and accept their regulations. After enactment by the states, business must file clean air source petitions which the state will then have three years to review and approve. Important pollutants, among others, are considered to be ethanol emissions from bakery stacks and vehicle exhaust, with the latter believed to account for 50 to 70% of all major air pollutants.

Wheat producers aver that if current herbicides are restricted, there will be a significant decrease in ton/acre wheat productivity. Flour millers aver that if current pesticides are restricted, there will be a significant increase in insect fragments in flour, compromising the cleanliness of the flour as specified by the U.S. Food and Drug Administration (FDA). Environmentalists argue that unless protection of the environment is given the highest priority, the future will be bleak for succeeding generations. In the meanwhile, there is scientific hope that pesticides and herbicides less damaging to the environment, but equally effective, will be developed. Also, that the new biotechnology advances will make possible development of wheat strains less dependent on protection from weeds and insect pests.

In 1992, a dramatic announcement was made that recombinant DNA techniques were used successfully to introduce a foreign gene into a common wheat strain. Previous recombinant DNA experiments with wheat failed because the technique for introducing the foreign gene damaged the wheat cell wall beyond the recovery point. However, coating tiny submicroscopic particles of tungsten with DNA from the foreign gene and injecting the DNA-coated particles into the wheat cell using a unique particle gun resulted in many new advances such as production of a new enzyme in the wheat which made it resistant to many powerful herbicides, even though previously these herbicides had retarded growth of the wheat substantially. There is some concern that the FDA would require extensive safety testing on wheat produced by recombinant DNA technologies.

It is also not clear whether industry will find this an attractive area for investment of research dollars, unless a way can be found to prevent genetically modified wheat from breeding true, and thereby establish a mandatory requirement for annual purchase of seed by wheat producers. This will, of course, increase the price of wheat, but may bring extraordinary benefits.

Although wheat is one of the earliest agricultural triumphs and has been basic in the human dietary for 8,000–10,000 years, it is exciting to view the prospects ahead. The worldwide wheat industry now has the opportunity to provide more than half of the caloric requirements in the "healthy" daily dietary, supplemented only by small amounts of vegetables and fruits, and still smaller amounts of dairy products and meat, poultry, fish, dry beans, eggs and nuts, with fats, oils and sweets to be used only sparingly. With new biotechnology opportunities, we can finally face the opportunity of overcoming worldwide hunger and make a major contribution to worldwide, global good health.

References

Anonymous. 1994. World Grain Vol. 11, No. 10, p. 6-13, Sosland Publishing Co., Kansas City, MO 64112, U.S.A.

Anonymous. 1993. World Grain, October 1993, p. 31, Sosland Publishing Co., Kansas City, MO 64112, U.S.A.

Fink, S. and Disbro, E. L. 1994, Independent Bakers Association Bulletin, February 10, 1994, Washington, DC 20007 U.S.A.

Jacob, H. E. 1994. Six Thousand Years of Bread, Its Holy and Unholy History, 1944, Doubleday and Co. Inc., New York, NY, reprinted 1970 Greenwood Press, Westport, CT, pp. 1-17.

Orth, R. A. and Shellenberger, J. A. 1988. Origin, Production, and Utilization of Wheat, in Wheat Chemistry and Technology, Vol I, Y. Pomeranz (Ed.) American Association of Cereal Chemists, St. Paul, MN, pp. 1-14.

Percival. J. 1921, The Wheat Plant, Dutton, New York, NY, quoted in Orth and Shellenberger (1988).

Contributors

Elsa de Sá Souza, Morixe Hnos. SACI. Cucha, Cucha 234, 1405 Buenos Aires, Argentina

Hamed Faridi, Nabisco Foods Group, 200 DeForest Avenue, East Hanover, NJ 07936

Jon M. Faubion, Department of Grain Science & Industry, Kansas State University, Manhattan, KS 66506

Kjell M. Fjell, Statkorn, P.O. Box 1367-Vika, 0114 Oslo 1, Norway

John T. Gould, Goodman Fielder Baking and Milling Co., 7 Barley Road, Leichhardt, NSW 2040, Australia

Simon S. Jackel, Plymouth Technical Services, 4523 Bardsdale Drive, Palm Harbor, FL 34685

Radomir Lásztity, Department of Biochemistry and Food Technology, Technical University of Budapest, 22 Lagymanyosi, Budapest, 1111, Hungary

Graham J. McMaster, Bread Research Institute of Australia, P.O. Box 7, North Ryde, 2113, Australia

Seiichi Nagao, Nisshin Flour Milling Company, Research Center, 5-3-1 Tsurugaoka, Saitama, 356 Japan

Roberto J. Peña, CIMMYT, Lisboa 27, Apdo. Postal 6-641, Mexico D.F., 06600, Mexico

Jalal Qarooni, Department of Grain Science and Industry, Kansas State University, Manhattan, KS, 66506

Philip G. Randall, Wheat Board, P.O. Box 908, Pretoria, 0001, South Africa

José L. Robutti, INTA, C.C31, 2700 Pergamino, Argentina

Hannu Salovaara, Dept. of Food Technology, University of Helsinki, P.O. Box 27 (Viikkib), Helsinki, FIN-00014, Finland

Wilfried Seibel, Federal Research Center for Cereal and Potato Processing, P.O. Box 23, D-4930, Detmold, Germany

Jiwan S. Sidhu, Department of Food Science and Technology, Punjab Agricultural University, Ludhiana-141004, India

Hans H. Traut, Wheat Board, P.O. Box 908, Pretoria, 0001, South Africa

Arie Wessels, Wheat Board, P.O. Box 908, Pretoria, 0001, South Africa

CHAPTER 1

Wheat Usage
in North America

Hamed Faridi and Jon M. Faubion

Introduction

Most of the wheat currently grown in North America consists of three species of the genus *Triticum*. *T. aestivum* L., a hexaploid, dominates world production. Hard Red Winter and Hard Red Spring classes are used primarily for leavened bread production. Soft Red Winter and Common White wheat classes are used primarily for pastries, crackers, and cookies. *T. compactum* Host, a club wheat, also a hexaploid type, is used for pastries in a similar manner as the Soft Red Winter wheats. Durum wheat, *T. durum* Dest., is a tetraploid species with extremely hard grain which is used for macaroni, spaghetti, and other pasta products.

To facilitate marketing, U.S. plant breeders have kept distinctive combinations of grain protein and hardness. The common wheat can be produced with a wide combination of hardness and protein. However, to complement their normal end uses, bread wheats are selected to have a desired combination of hardness, protein level, and protein quality. Soft wheat hexaploids are selected to have soft endosperm texture and low protein. Production in areas of high rainfall helps to promote the soft texture and low protein content.

The tetraploids (e.g. durum) are reported to be associated with hard grain types. Diploids have soft grains and hexaploids can be variable in kernel hardness. The club wheats, as hexaploids, could be selected as hard types, but have been

1

maintained as a soft market class (Anon, 1992).

Wheat Standards and Grades. As the grain trade developed in the United States, there was a need for a reliable commercial language to allow buyers and sellers at separate locations to trade without an exchange of samples. Grain trade organizations took the lead in setting standards but could never agree on consistent standards. The demand for uniform grades and inspection resulted in 26 different bills between 1903 and 1916 in the United States, which provided for federal supervision of grain grading or for outright federal grain inspection (USDA, 1957; Mattern, 1991). These bills and hearings produced the U.S. Grain Standards Act, which was passed August 11, 1916. The act provided in part for (a) the establishment of official U.S. grain standards, (b) the federal licensing and supervision of the work of grain inspectors and, (c) establishment of an Appeals Board for complaints on grades assigned by the licensed inspectors. Currently, U.S. standards are in effect for wheat, corn, barley, oats, rye, triticale, sorghum, flaxseed, soybeans, and mixed grain (USDA, 1988).

According to the U.S. Standards for Wheat, wheat is defined as grain that, before the removal of dockage, consists of 50 percent or more common wheat (*Triticum aestivum* L.), club wheat (*T. compactum* Host.), and durum wheat (*T. durum* Desf.) and not more than 10 percent of other grains for which standards have been established under the United State Grain Standards Act and that, after the removal of the dockage, contains 50 percent or more of whole kernels of one or more of these wheats (Federal Register, 1987).

Wheat is divided into seven classes based on color, kernel and varietal characteristics. The seven classes are: Hard Red Spring wheat, Hard Red Winter wheat, Soft Red Winter wheat, Durum wheat, White wheat, Unclassed wheat, and Mixed wheat. The classes Hard Red Spring wheat, Durum wheat, and White wheat are further divided into subclasses. Each class and subclass is divided into five U.S. numerical grades and U.S. Sample grades (Table 1-1). Special grades are provided to emphasize special qualities or conditions affecting the value of wheat and are added to and make a part of the grade designation. Special grades do not affect the numerical or sample grade designation.

Hard Red Spring wheat (HRSW) is divided into the following three subclasses: a) Dark Northern Spring wheat: HRSW with 75% or more of dark, hard, and vitreous kernels. b) Northern Spring wheat HRSW with 25% or more, but less than 75%

Table 1-1. Official U.S. Grades and Grade Requirements

Grade	Minimum Limits — Test Weight per Bushel		Maximum Limits of						
	Hard Red Spring wheat of White Club wheat[a] (lb)	All other classes and subclasses (lb)	Damaged Kernels — Heat-Damaged Kernels (%)	Damaged Kernels — Total[b] (%)	Foreign Material (%)	Shrunken and Broken Kernels (%)	Defects[c] (%)	Wheat of Other Classes[d] — Contrasting Classes (%)	Wheat of Other Classes[d] — Total[e] (%)
U.S. No. 1	58.0	60.0	0.2	2.0	0.5	3.0	3.0	1.0	3.0
U.S. No. 2	57.0	58.0	0.2	4.0	1.0	5.0	5.0	2.0	5.0
U.S. No. 3	55.0	56.0	0.5	7.0	2.0	8.0	8.0	3.0	10.0
U.S. No. 4	53.0	54.0	1.0	10.0	3.0	12.0	12.0	10.0	10.0
U.S. No. 5	50.0	51.0	3.0	15.0	5.0	20.0	20.0	10.0	10.0

U.S. Sample grade is wheat that:
1) Does not meet the requirements for the grades U.S. Nos. 1, 2, 3, 4, or 5; or
2) Contains 32 or more insect-damaged kernels per 100 grams of wheat; or
3) Contains 8 or more stones or any number of stones which have an aggregate weight in excess of 0.2 percent of the sample weight, 2 or more pieces of glass, 3 or more crotalaria seeds (*Crotalaria* spp.), 2 or more castor beans (*Ricinus communis* L.), 4 or more particles of an unknown foreign substance(s) or a commonly recognized harmful or toxic substance(s), 2 or more rodent pellets, bird droppings, or equivalent quantity of other animal filth per 1,000 grams of wheat; or
4) Has a musty, sour, or commercially objectionable foreign odor (except smut or garlic odor); or
5) Is heating or otherwise of distinctly low quality.

[a] These requirements also apply when Hard Red Spring or White Club wheat predominate in a sample of Mixed wheat.
[b] Includes heat-damaged kernels.
[c] Defects include damaged kernels (total) foreign material, and shrunken and broken kernels. The sum of these three factors may not exceed the limit of defects for each numerical grade.
[d] Unclassed wheat of any grade may contain not more than 10.0 percent of wheat of other classes.
[e] Includes contrasting classes.

Source: Federal Register, 1987

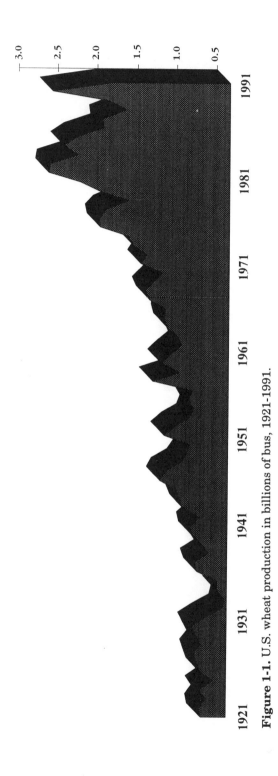

Figure 1-1. U.S. wheat production in billions of bus, 1921-1991.

of dark, hard, and vitreous kernels. c) Red Spring wheat: HRSW with less than 25% of dark, hard, and vitreous kernels.

Durum wheat is divided into the following three subclasses: a) Hard Amber Durum wheat: Durum wheat with 75% or more hard and vitreous kernels of amber color. b) Amber Durum wheat with 60% or more, but less than 75% of hard and vitreous kernels of amber color. c) Durum wheat: Durum wheat with less than 60% of hard and vitreous kernels of amber color.

All varieties of white wheat are divided into the following four subclasses. a) Hard White wheat: White wheat with 75% or more of hard kernels. It may contain not more than 10% of White Club wheat. b) Soft White wheat: White wheat with less than 75% of hard kernels. It may contain not more than 10% of White Club wheat. c) White Club wheat: White Club wheat containing not more than 10% of other White wheat. d) Western wheat: White wheat containing more than 10% of White Club wheat and more than 10% of other White wheat.

Unclassed wheat is any variety not classed under the standards. It includes Red Durum wheat and wheats of other colors. Mixed wheat consists of any mixture of wheat that consists of less than 90% of one class and more than 10% of one other class, or a combination of classes meeting the definition of wheat (Miller and Kirk, 1987; Mattern, 1991).

Farm Output. The ability of wheat farmers to respond to any expansion that might develop in export demand and to greater domestic requirements hinges in large measure on decisions made by the U.S. government. Government is now the largest single factor in the United States grain industry, in both production and marketing of wheat. On the supply side, the government imposes acreage reduction requirements on growers for participation in price support loan programs. On the demand side, the government firmly controls exports by determining eligibility of importing countries for subsidies (Anon, 1992).

In 1922–23, the yield per acre in the United States was 13.8 bus. For the 1990–91 crop year, the per-acre yield was triple that amount, at a record 39.5 bus. That performance is the result of the development, starting in the late 1960s, of new cultivars with high-yield capability.

The record in U.S. wheat production was attained in 1981 at 2,785,400,000 bus. The second largest output was in the following year, 1982 at 2,764,967,000 bus. In 1915, wheat production for the first time exceeded 1 billion bus., at 1,008,600,000 bus. (Fig. 1-1).

The average wheat grower owns 151 acres of farmland and derives most of his income from farming, according to the 1987 Census of Agriculture issued by the Economic Research Service (ERS) of the U.S. Department of Agriculture. Of course, variations exist. For example, the average size of a Montana wheat farm was 453 acres in 1987, while the average Ohio and Indiana wheat farm was only 32 acres.

Flour Milling in the U.S.

Flour milling could hardly be more favorably positioned than at present, reflecting the astounding growth in consumption and a continuing favorable balance between demand and capacity. Flour production is near record levels, spurred by a rising trend in consumption that got under way in the early 1970s after reversing the downward pattern that had prevailed since the turn of the century. The average rate of milling operations in 1991 was about 90% of capacity, based on a six-day week (Anon, 1992).

The driving force in bringing milling capacity into line with demand has been a revolutionary shift from animal to crop foods as the mainstay of the American diet. Per capita consumption of flour, including semolina, in 1991 was 136 lbs, according to data provided by the ERS. The preliminary estimate for 1991 is unchanged from 136 lbs as the estimate for 1990, but above 129 lbs in 1989. The estimates for 1990 and 1991 mark a continuation of the rising trend from the all-time low of 110 lbs in 1972. Prior to 1972, per capita consumption had fallen steadily from 217 lbs in 1909, the first estimate issued by ERS (see Fig. 1-2).

Leading the growth in percentage increase in per capita flour consumption has been semolina. The ERS estimates that semolina and durum flour consumption in 1990 at 10.5 lbs, is up 52% from 6.9 lbs in 1970. That compares with wheat flour, excluding durum, at 124.6 lbs in 1990, up 20% from 104 lbs in 1970. Even that performance in semolina consumption (as measured by ERS) falls far short of estimates by the National Pasta Association that per capita pasta consumption in the U.S. has climbed to around 18 lbs. Because a hundred-weight of semolina produces about 92 lbs of pasta, the disparity between the two sources is substantial. Blending of semolina or durum flour with hard wheat farina or flour could account for a portion of that difference.

At the turn of the century, home baking accounted for a

great percentage of flour consumption, with commercial bakeries accounting for only 10% of the total, against 90% as family flour.

By 1945, the bakery portion had climbed to 60% as home baking diminished with the entry of women into the workforce. By 1963, commercial bakers accounted for 75% of flour usage. The ratio of flour usage by bakers and households had come full circle by 1990, with much less than 10% of the total as family flour and consumer mixes and the balance as bakery flour.

Daily milling capacity of wheat flour, including semolina and durum flour, in 1992 was 1,354,168 cwts, compared with 1,014,427 cwts in 1970, an increase of 33%. The most prominent development in milling capacity has been in the durum portion of the wheat flour total. Semolina and durum flour milling capacity currently is at a record 133,595 cwts, nearly triple the 1970 total of 51,678 cwts with another 19,000 cwts of durum capacity is under construction.

At the end of World War I, there were more than 2,000 flour mills in the U.S. By 1970, the number had dropped to 358 mills with a total capacity of 1,014,427 cwts, including 51,678 cwts of durum. In 1982, the number was down to 262 mills with capacity of 1,152,071 cwts, including 81,400 cwts of durum. Currently 226 mills account for capacity of 1,354,168 cwts, including 133,595 cwts of durum (Fig. 1-3).

As the number of flour mills in the U.S. continues in a decreasing trend, the average size of individual plants is increasing. The 1992 Milling Directory of *Milling and Baking News*

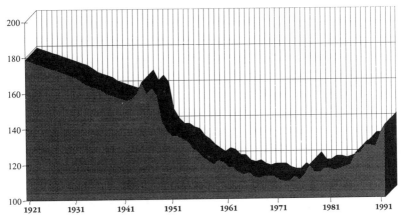

Figure 1-2. Per capita consumption in lbs, 1922-1991.

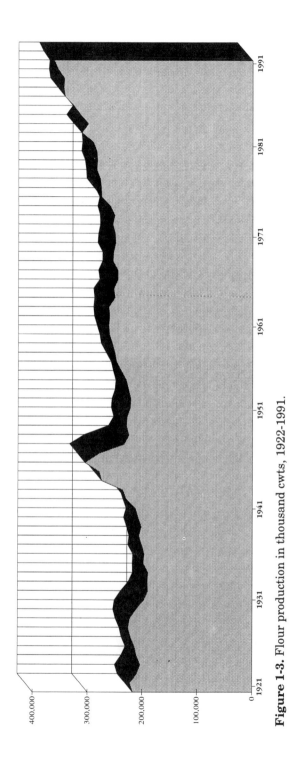

Figure 1-3. Flour production in thousand cwts, 1922-1991.

lists 44 wheat flour mills, excluding durum plants, with a daily capacity of 10,000 cwts or more, for a total of 669,100 cwts in that category. Fifty-five mills are listed with capacity of 5,000 to 9,999 cwts, for a total of 382,400 cwts, along with 58 with capacity of 1,000 to 4,999 cwts, for a total of 154,020 cwts.

The 10 leading milling companies have 1,079,140 cwts of wheat-durum-rye daily capacity. This represents 80% of the industry's 1,354,160-cwt total. In 1982, the 10 largest companies accounted for 800,900 cwts, or 70% of the 1,152,071-cwt total. In 1970, the 10 largest millers accounted for 565,170 cwts of capacity, or 56% of that year's total capacity of 1,014,417 cwts (Anon, 1992).

Consumption of Baked Products in the U.S.

Per capita consumption of bread, cake and related items produced by wholesale bakers will increase at an annual rate of 1.7% in the 1992–97 period, according to the "1992 Industrial Outlook" of the U.S. Department of Commerce. The Outlook, which provides an annual update for 350 manufacturing and services industries, states that per capita consumption of all wholesale-produced baked foods will rise 1.6% annually over the next five years.

"In 1991, per capita consumption of all bread increased 1.3%," the Outlook says. "After 26 years of decline, consumption of white bread began to rise in 1984. This trend has continued due to lower retail prices, improving quality, more effective marketing, and better packaging. Variety bread also gained ground, though at a slower pace than in earlier years." The Outlook places per capita consumption of bread in 1991 at 50.50 lbs, compared with 49.87 lbs in 1990. It forecasts all-bread consumption at 51.19 lbs in 1992 and at 52.01 lbs in 1993.

The strong performance of white bread in recent years is reflected in a revision in the 1992 Outlook of a forecast made a year earlier which predicted that per capita consumption of variety bread in 1992 would outstrip white bread for the first time. The Department's new Outlook forecast 1992 per capita consumption of variety bread at 22.78 lbs, compared with 28.42 lbs for white bread. For 1993, the Outlook forecasts per capita variety bread consumption at 23.28 lbs and white bread at 28.73 lbs. The variety bread consumption forecast for 1992 included 6.85 lbs of hearth-type bread; 10.01 lbs of whole-wheat and cracked-wheat loaves; 2.18 lbs of rye-bread; and

3.74 lbs of all other variety bread (Tables 1-2 and 1-3).

While the severity of the 1992 recession's impact on baking and pasta manufacturers varied from region to region and even in individual markets within regions, overall unit sales of bread as 1992 began were running about 1% below the prior year. Similar decreases were noted in cookies and crackers. Cookie volume was off 1% from the previous year and cracker volume was flat. Snack cake sales, the purchase of which is highly discretionary in nature, have been flat, at best. Although pasta enjoyed a 3% increase in per capita consumption in 1991, manufacturers noted consumer resistance to higher prices as well as a marked increase in imports (Anon, 1992). During this period, bakers began looking to "non-traditional" outlets in order to expand volume. Several have established separate sales and distribution operations to reach these "alternative format" stores, including warehouse clubs, deep-discount drug stores and other mass merchandisers.

The 1985–86 period was marked by the most significant plant closings in wholesale baking since the early 1970s, when baking was caught in a cost-price control squeeze that left many bakers with no option but to shut down. Most of the closings of the mid-1980s were not forced by factors outside of baking's control but were strategic management decisions made to improve short- and long-term profitability. In March 1985, for example, the five largest bread and cake baking companies were operating a total of 194 baking plants. Eighteen months later, that number had dipped to 175.

The pasta industry in the 1980s changed from a business that was largely local in character and family-owned to one where the business is dominated by companies from outside the industry, companies whose operations previously focused on chocolate, on dairy products, and on soft drinks. By 1990, Borden, Hershey, and CPC held a combined market share of 66% of North American branded dry pasta sales.

Acquisition activity in baking continued at a rapid pace in the late 1980s and into the 1990s. Some companies and plants changed owners more than once in a five-year period. The implementation of the U.S.-Canada Free Trade Agreement in 1989 prompted additional restructuring in U.S. and Canadian baking through acquisitions, joint ventures and market agreements. Similar investments are being made by companies on both sides of the U.S.-Mexico border in light of the successful enactment of the North American Free Trade Agreement (NAFTA).

Table 1-2. Per Capita Consumption of Bread and Related Products in lbs.

Year	1988	1989	1990	1991*	1992*	1993+	1994	1995	1996	1997	1998	1999	2000	2001
All Bread	48.68	49.28	49.87	50.50	51.19	52.01	52.85	53.78	54.83	55.88	56.96	58.20	59.54	60.64
White	27.77	27.80	27.92	28.19	28.41	28.73	29.07	29.47	29.95	30.35	30.83	31.42	32.08	32.44
Variety	20.91	21.48	21.95	22.31	22.78	23.28	23.79	24.32	24.88	25.52	26.13	26.78	27.46	28.19
Hearth	6.06	6.29	6.49	6.66	6.85	7.08	na	na	na	na	na	na	na	na
Whole Wheat, cracked	9.55	9.70	9.80	9.88	10.01	10.13	na	na	na	na	na	na	na	na
Rye	2.03	2.08	2.12	2.14	2.18	2.21	na	na	na	na	na	na	na	na
All other	3.27	3.41	3.54	3.63	3.74	3.86	na	na	na	na	na	na	na	na
Rolls	21.70	22.31	22.81	23.09	23.47	23.91	24.33	24.70	25.08	25.35	25.63	25.91	26.20	26.72
Hamburger, hot dog	13.05	13.20	13.30	13.33	13.40	13.46	na	na	na	na	na	na	na	na
Bagels	2.47	2.75	2.99	3.15	3.33	3.56	na	na	na	na	na	na	na	na
Brown'n serve	1.32	1.34	1.35	1.35	1.36	1.37	na	na	na	na	na	na	na	na
Hearth	1.34	1.36	1.38	1.39	1.40	1.41	na	na	na	na	na	na	na	na
English Muffins	1.70	1.69	1.68	1.65	1.64	1.63	na	na	na	na	na	na	na	na
Croissants	0.41	0.45	0.48	0.51	0.53	0.57	na	na	na	na	na	na	na	na
Other bread-type rolls	1.41	1.52	1.63	1.71	1.81	1.91	na	na	na	na	na	na	na	na
Sweet Yeast Goods	3.78	3.90	3.99	4.04	4.11	4.11	4.28	4.39	4.50	4.58	4.65	4.73	4.80	4.87
Donuts	1.29	1.41	1.50	1.56	1.63	1.71	1.81	1.90	2.01	2.10	2.20	2.29	2.37	2.45
All other	2.49	2.49	2.49	2.48	2.48	2.40	2.48	2.48	2.49	2.47	2.46	2.45	2.43	2.42
Soft Cakes	7.51	7.52	7.73	7.90	8.13	8.34	8.55	8.72	8.89	9.02	9.17	9.32	9.49	9.66
Snack cakes	6.05	6.38	6.64	6.86	7.13	7.36	7.60	7.79	7.97	8.13	8.29	8.46	8.63	8.80
All other	1.16	1.14	1.09	1.04	1.00	0.98	0.95	0.94	0.92	0.89	0.87	0.86	0.86	0.86
Pies	1.74	1.71	1.69	1.68	1.66	1.64	1.64	1.64	1.63	1.63	1.63	1.63	1.63	1.64
Snack pies	1.41	1.43	1.44	1.46	1.47	1.48	1.50	1.52	1.53	1.53	1.55	1.56	1.57	1.59
All other	0.33	0.28	0.25	0.22	0.19	0.16	0.14	0.12	0.11	0.09	0.08	0.07	0.06	0.05
Cake Type Donuts	1.22	1.31	1.22	1.09	0.99	0.92	0.86	0.82	0.80	0.80	0.79	0.79	0.78	0.78

Source: Anon, 1992

Table 1-3. Per Capita Consumption of Cookies and Crackers in lbs

Year	1988	1989	1990	1991	1992	1993	1994	1995	1996	1997	1998	1999	2000	2001
Cookies	12.23	12.91	12.58	12.15	12.19	12.29	12.56	12.88	13.16	13.37	13.53	13.69	13.85	14.02
Sandwich	2.96	3.14	3.15	2.99	3.02	3.06	3.09	3.12	3.16	3.19	3.22	3.25	3.29	3.32
Marshmallow	0.27	0.26	0.26	0.25	0.24	0.23	0.21	0.20	0.18	0.18	0.16	0.15	0.14	0.13
Wafers for ice cream sandwiches	0.29	0.31	0.31	0.30	0.31	0.31	na	na	na	na	na	na	na	na
All others	8.71	9.20	8.86	8.61	8.62	8.69	8.93	9.23	9.49	9.67	9.81	9.94	10.08	10.23
Crackers	7.97	7.96	8.08	8.11	8.15	8.23	8.39	8.61	8.85	9.01	8.31	7.58	6.85	6.13
Graham	0.70	0.69	0.68	0.67	0.67	0.66	0.66	0.66	0.66	0.66	0.66	0.66	0.66	0.66
Saltines	2.10	2.02	1.98	1.93	1.88	1.85	1.83	1.81	1.77	1.74	1.71	1.68	1.66	1.63
Cracker sandwich	0.52	0.54	0.59	0.67	0.73	0.76	0.53	0.53	0.54	0.54	0.48	0.41	0.35	0.29
Other	4.65	4.71	4.83	4.84	4.87	4.96	5.17	5.41	5.68	5.87	5.29	4.68	4.06	3.45
Pretzels	1.11	1.16	1.21	1.21	1.21	1.22	1.27	1.32	1.37	1.42	1.28	1.12	0.97	0.82

Source: For 1988-1993, Annual Survey of Manufacturers, U.S. Department of Commerce.
For 1994-1997, Annual Survey of Manufacturers (statistics include in the Survey but not presented in its tables.)
For 1988-2001, projections based on unpublished International Trade Administration data.
*Estimates by I.T.A.
+Forecasts by I.T.A.
na:Not Available
From: Anon, 1992

Market for U.S. Wheat

There was no clear trend in export demand for United States wheat in the 1980s. Uncertainty marks the future, both in the short term and long term. In contrast, domestic wheat consumption posted significant gains, led by the substantial increase in food use. It has become very clear that the U.S. baking industry represents the most consistent outlet for American wheat. For the 1990s and beyond, the outlook for the export wheat business is unclear, while no evidence has emerged to indicate a reversal in the rising trend in domestic food use.

From 1970 to 1991, annual disappearance of U.S. wheat for food use increased 50%, from 517 million bus to 775 million bus. Utilization as animal feed, seed and residual varied from year to year. Per capita consumption of flour reached its nadir of 110 lbs in 1972, a year that marked the start of the climbing trend, reaching 136 lbs in 1992 (Fig. 1-4).

Recent annual variation in the volume of world wheat trade most often has reflected fluctuations in export shipments to what was the Soviet Union and to China. The loss of U.S. share of world trade in wheat and flour has been the direct result of intense competition with the European Community for markets. When the U.S.S.R. made its first large purchases of wheat in 1972–73, the U.S. share of the world wheat trade reached 46%, against 18% for the E.C. Through the rest of the 1970s, the U.S. share of world wheat exports ranged from 41% to 53%, while the E.C. portion ranged from 17% to 22%. Through its system of restitutions, actually export subsidies, the E.C. in the 1980s boosted its share, including intra-Community transactions, to as high as 34% of the global total, surpassing even the U.S. share in several crop years (Fig. 1-5).

With the recent slippage in U.S. exports and rise in domestic use, greater importance is shifting to the milling and baking quality of wheat. However, a substantial and expanded export business is necessary to assure that farmers will grow sufficient quantities of the classes and qualities of wheat needed by breadstuffs manufacturers. Already under way and certain to continue into the 1990s and beyond are efforts by the baking industry to promote the production of wheat with desirable, even superior, baking characteristics. These efforts include contracting with farmers to grow specific cultivars and with the grain industry to provide identity preservation of the types of wheat needed (Anon, 1992).

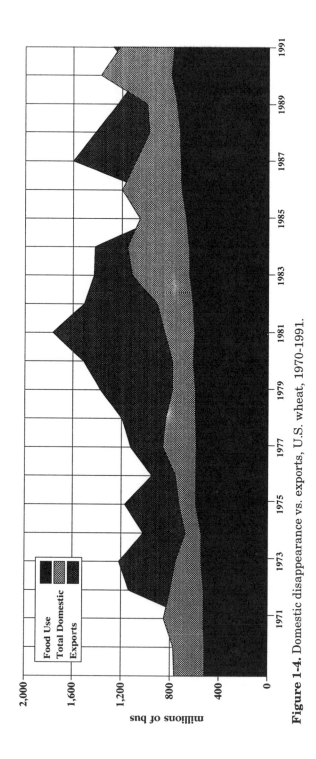

Figure 1-4. Domestic disappearance vs. exports, U.S. wheat, 1970-1991.

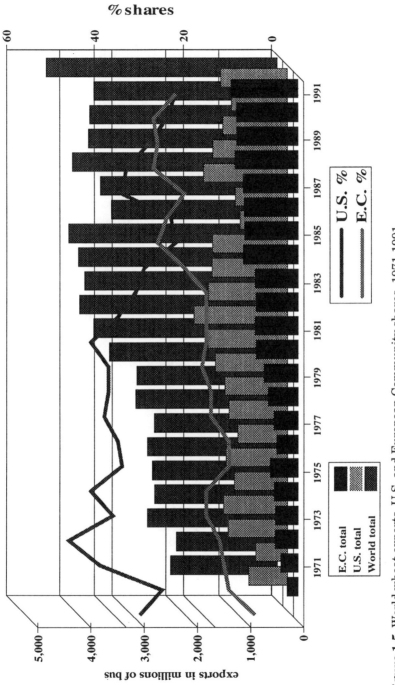

Figure 1-5. World wheat exports, U.S. and European Community shares, 1971-1991.

Farm Output in Canada

Canada, a major wheat producer and exporter similar to the U.S. has demonstrated the ability to increase yield per acre as well as total farm output. The production of wheat has doubled from 1968 to 1992. Domestic consumption as well as exports have increased accordingly. Because the spectrum of baked products made in Canada is nearly identical to that of the United States, there will be no separate discussion focused on products made in the former.

Hard Wheat Products

Manufacturing of Bakery Foods. The consolidation of the wholesale industry into large corporations continues. Although fully baked, wholesale bakery products remain the major component of the industry, there is a significant trend to splitting production between a central plant (generally a wholesale bakery) and a point of distribution (a supermarket). In the central plants, the doughs for breads and yeast-leavened products are mixed, usually by a no-time dough process (illustrated in Table 1-4), then frozen. The frozen dough is transported in refrigerated trucks to the point of distribution (the "in-store bakery") where the doughs may be kept frozen and processed immediately or within 2 weeks. The processing (bake-off), includes defrosting of doughs, proofing, and baking. These doughs may also be marketed in a frozen state directly to the consumer. When manufactured for that purpose, the dough formulas contain higher than normal amounts of yeast (6%, flour basis), high levels of oxidants, and dough strengtheners to enhance the stability of doughs over its 12–18 weeks of frozen storage. The advantage of baking yeast-leavened products in in-store bakeries is the flavor and high quality of texture of the fully baked products. Although these attributes are attractive to the consumer, products prepared by this type of operation require a higher price tag than breads from wholesale production. Consequently, the products from in-store bakeries are generally premium items while the wholesale fully baked products are for general consumption (Kulp, 1991).

Commercial Production Methods. The flow chart of the main bread manufacturing processes currently in use are shown in Figures 1-6 to 1-8. Figures 1-6 and 1-7 depict the predominant manufacturing processes: the sponge and dough

Table 1-4. White Pan Bread

Ingredient	Sponge Dough 2		Straight Dough (%)	Straight No-time Dough (%)	Brew Bread (%)
	Sponge (%)	Dough (%)			
Flour[a]	70.0	30.0	100.0	100.0	100.0
Brew	---	---	---	---	35.0
Water	40.0	24.0	64.0	65.0	32.0
Yeast, compressed	3.0	---	2.5	3.5	---
Salt	---	2.0	2.0	2.0	1.4
Sugar or sweetener solids	---	8.5	8.0	6.0	7.0
Shortening	---	3.0	2.75	2.75	2.75
Yeast food	0.5	---	0.5	0.6	0.5
Nonfat dry milk or milk replacer	---	2.0	2.0	2.0	2.0
Fungal protease	0.5	---	0.25	0.5	---
L-Cysteine, ppm	---	---	---	40.0	---
Potassium Bromate, ppm	---	---	15.0	30.0	30.0
Ascorbic Acid, ppm	---	---	---	60.0	---
Vinegar (100 grain)	---	---	---	0.5	---
Monocalcium phosphate	---	---	---	0.25	---
Mono- and diglycerides, hydrate	---	0.5	0.5	0.75	0.5
Dough strengtheners	---	---	---	---	0.2
Calcium propionate	---	0.2	0.2	0.2	0.2

[a]About 11.2-11.7% protein, 0.44-0.46 ash, enriched according to U.S. standards (14% m.b.). All ingredients are given in bakers % (flour = 100%).

Procedures

1) Sponge and Dough: Sponge: Sponge/dough ratio: 70/30; mixing: 1 min low, 3 min high speeds (77°F/25°C); fermentation: 3-5 hr (86°F/30°C). Dough Mixing: 1 min low, 10-12 min high (80°F/27°C); rest period: 15-20 min. Divide into 18 oz (551g) pieces for 1 lb (454g) loaves. Round and give 7 min rest period, sheet, shape, and pan; proof: 55 min.[(107°F/42°C)/85% rel. humidity]; bake: 18 min at 445°F (230°C).

2) Straight Dough: One-step process. Mixing: 1 min at low and 15-20 min at high speeds (78-82°F/26.5-27.5°C). Fermentation time 2 hr at 86-95°F (30-35°C)/85% relative humidity.
This procedure is used by retailers for speciality breads.

3) Straight Dough - No-time: Mix 1 min at low and 10-15 min (80-84°F/27-29°C). Proof for 55 min [(107°F/42°C/85% rel. humidity].

4) Brew Procedure: Brew: Disperse ingredients by high speed agitation (5 min, 80°F/28°C). Fermentation (low agitation) for 1.5 hr (88-92°F/31-33°C). Add to dough ingredients and proceed as in sponge and dough process. Brew consists of 80.95% water, 7.75 % sweetener solids, 8.0% compressed yeast, 3.8% salt, and 0.2-0.5% buffer (calcium carbonate plus ammonium sulfate); 35% of this brew is added to the dough.

Source: Kulp and Dubois, 1982

and liquid ferment methods. Sponge and dough bread production predominates in the United States, accounting for about 61% of total bread production. Other processes used are the straight dough, continuous, no-time, and Chorleywood procedures. Of the remaining four processes, the straight dough method is mainly used in retail operations and in production of specialty breads. The continuous process lost its popularity due to product quality issues. The no-time process is a minor production method, except for manufacturing of some specialty breads, and the Chorleywood processes is common in England and many countries throughout the world but not in North America. A comparison of production steps in these methods is given in Figure 1-8, showing the typical steps and processing times (Kulp, 1991).

Production of Buns and Rolls. Rolls and buns (hamburger, hot dog rolls) form an important specialty segment of the hard wheat flour market. The products require not only strong hard wheat flour of high protein content, but also, occasionally supplementation of vital wheat gluten. Their production involves mixing, dividing, rounding, proofing, molding and panning operations, which are generally performed by integrated units that incorporate the various equipment sections needed to produce the panned dough pieces.

Sweet Dough. The production of sweet dough products, e.g., Danish pastry, sweet rolls, coffee cakes, etc., is done either

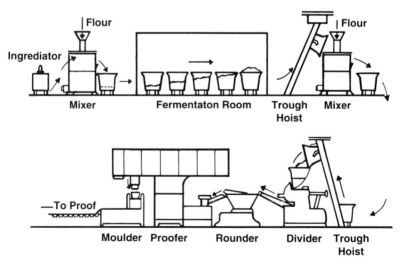

Figure 1-6. Sponge and dough bread production line. Source: Semling 1988.

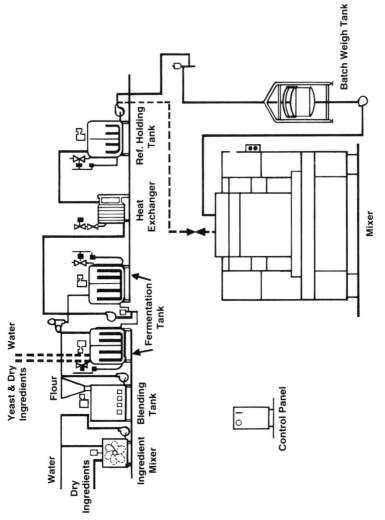

Figure 1-7. Flow chart for liquid ferment process. Source: Kulp, 1983.

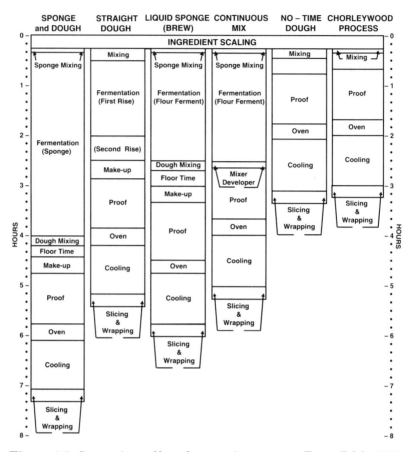

Figure 1-8. Comparison of bread-processing systems. From: Zelch 1988

manually or by mechanical devices which perform most of the basic operations.

Types of Breads and Typical Formulas

White Pan Bread. The major type of bread manufactured in the United States is white pan bread. This type of bread is a standardized product and is regulated in respect to formulation, final moisture (38% maximum) and enrichment. A typical formula for production of white pan bread by major processes is shown in Table 1-4 along with the basic procedures used in its production. The formula for continuous process bread is slightly different (Table 1-5).

Table 1-5. Continuous Bread Making Formula - Nonflour Brew

Ingredients	(%)[a]	Dough Composition	
Water	60.32	Flour, %	100
Sugar	7.61	Brew, %	74
Yeast	2.67	Shortening blend, %	3
Yeast food	0.50	Water, %	2
Salt	2.10	$KBrO_3$ ppm	60
Milk Powder	0.33	KIO_3 ppm	12
Mold Inhibitor[b]	0.10		

[a]Total flour basis
[b]Calcium propionate

Source: Redfern et al., 1964

White Specialty Breads. Breads of this group are called in the market; homestyle, farm, country, or range breads. They differ in general, from the conventional white pan bread in internal grain and texture characteristics. The grain is more open and their texture more coarse. These properties, reminiscent of home-made breads, are achieved primarily by not developing the doughs fully (undermixing). A typical formula is shown in Table 1-6. Flour used in these breads is unbleached to obtain an off-white natural crumb color. Also, relatively low sugar in the formula is common for these breads. *Premium white bread* is a dense bread usually produced by a straight dough process as described in Table 1-4. *White hearth breads* are produced with or without lactic acid ferment, called a sour fermentation by the trade. The main difference between pan and hearth breads is in the baking step. Hearth breads are baked directly on an open hearth, whereas the others are

Table 1-6. Home-Style Bread Formula[a]

Ingredients	Total (%)	Sponge (5%)	Dough (%)
Flour, unbleached	100	65	35
Water (variable)	65	40	25
Corn syrup	3	--	3
Salt	2	--	2
Shortening	3	--	3
Yeast	2	2	--
Yeast food	0.375	0.375	--
Rolled wheat	2	2	--
Mold inhibitor	as needed	2	as needed

[a] Ingredients based on 100 parts flour. Sponge temperature, 67°F. Sponge fermentation, 4 hr. Dough temperature, 82°F. Floor time, 15-20 min. Give medium proof. Bake 420-430°F., 30 min.

Source: Ponte, 1981

baked in pans. The type of heat transfer during baking leads to formation of solid, crisp, flavorful crust, and other attributes associated with the type of breads. Baking often includes the injection of live steam into the oven to prompt the crust formation. Sour hearth breads are a variation on the hearth bread theme. The sour that is used in that type of bread is a flour dough (generally rye) fermented by lactic and acetic acid bacteria. It is prepared in the bakeries by natural fermentation or purchased in dry form. San Francisco hearth bread is a special sour dough all wheat flour bread in which the sour consists of a mixture of a special strain of yeast (*Torulopsis holmii*) and lactobacillus (*Lactobacillus sanfranciscus*). The optimum final pH of the dough is quite low (~3.6–3.7) which results in a dough with increased flow properties (slack).

Wheat Breads. The most common variety of this group is "wheat" bread, which is made from a flour blend of 20–30% whole wheat flour and 80–70% patent flour. Whole wheat bread is a standardized product and must be produced from whole wheat flour only (Kulp, 1991).

Rye and Pumpernickel Bread. Rye breads do not enjoy the popularity in the United States that they enjoy in Northern Europe. They are produced by a number of different formulas depending on local ethnic preferences, e.g., American, Jewish, German, Swedish, Black Forest, Bohemian. The basic flour requirement is a blend of rye flour (white and/or medium) and strong patent or clear wheat flours. Addition of vital wheat gluten to the dough is generally necessary to strengthen it.

Mixed Grain Breads. Mixed grain breads are made with wheat flour and other grains, as well as vegetable materials. A great variety of grains and vegetables have been used including corn, flax, millet, triticale, buckwheat, barley, oats, alfalfa, soy, potato, rye, rice, and sauerkraut. These materials can and are used as flours, grits, or as whole grains (Kulp, 1991).

Soft Wheat Products

Crackers. Crackers contain little or no sugar and moderate levels of fat (Hoseney, 1986; Manley, 1983). The dough generally contains low levels of water. There are three major types of crackers: saltine, chemically leavened, and savory. Crackers, particularly saltines, are deceptively simple food systems. In fact, the process required to produce acceptable saltines is both lengthy and complex. A brief description of the process follows. Saltine crackers are made by a sponge and

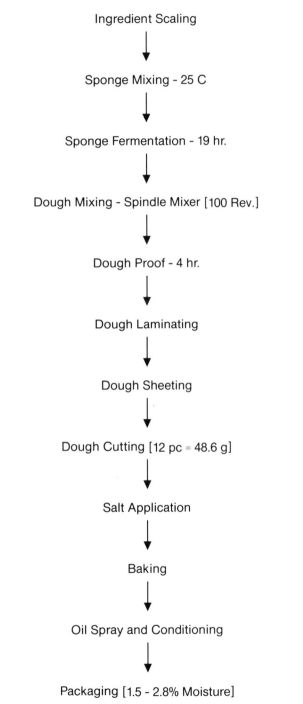

Ingredient Scaling

↓

Sponge Mixing - 25 C

↓

Sponge Fermentation - 19 hr.

↓

Dough Mixing - Spindle Mixer [100 Rev.]

↓

Dough Proof - 4 hr.

↓

Dough Laminating

↓

Dough Sheeting

↓

Dough Cutting [12 pc = 48.6 g]

↓

Salt Application

↓

Baking

↓

Oil Spray and Conditioning

↓

Packaging [1.5 - 2.8% Moisture]

Figure 1-9. Saltine cracker production process. Source: Hoseney, 1988.

dough process that requires about 24 hours, much of it for fermentation (Fig. 1-9). Because cracker sponges are mixed just long enough to wet the flour, gluten development occurs only to a limited extent at this stage. During the next 19 hours of sponge fermentation, the consistency of the sponge changes drastically as the sponge becomes more acidic and less elastic.

After fermentation, the sponge is mixed with other dough ingredients and the "dough-up" flour; the dough is allowed to relax for 4–6 hours. After the relaxation/resting period the dough is taken to the hopper of the sheeter. Sheeting this extensible, non elastic dough thru multiple sets of rolls produces the thin dough piece required in the next step.

Cracker doughs are laminated after exiting the sheeting rolls. After lamination, multiple pairs of heavy steel rolls (called gauge rolls) gradually reduce the dough sheet thickness to that desired for cutting. Typically, there are two or three pairs, although only one pair may be used for short doughs and more than three may be used where gentle reductions in dough thickness are necessary. As a rule of thumb, the reduction in thickness should be about 2:1 for each pass through a roll pair, although ratios of up to 4:1 are used. Obviously, the greater the ratio, the more work and stress is put into the dough and the more its physical properties change (Matz and Matz, 1978, Manley, 1983).

Chemically leavened crackers are generally called "snack crackers" and have a final pH of about 6.5 as compared to 7.0–7.1 for saltines. After a single stage mixing and a relatively short rest time (2–4 hrs), snack cracker dough is sheeted to form a continuous ribbon which is then laminated, with a light application of dusting flour between the layers. Graham crackers are chemically leavened semi-sweet crackers in which part of the white flour (10–40%) is replaced by whole wheat flour. Figure 1-10 outlines the process for production of chemically leavened crackers.

Flavored or savory crackers are well accepted in the U.S. market and account for much of the recent growth in cracker sales. The intense savory flavors are produced by adding the appropriate flavoring agents directly to the dough before or to the surface of the crackers after baking. Savory or cheese crackers are generally produced from fermented doughs. The yeast fermentation and the lower pH improve the cheese flavor. The formulation and processing is similar to that of soda crackers, with adjustments to compensate for the increased fat and moisture content of the cheese (Hoseney et al., 1988).

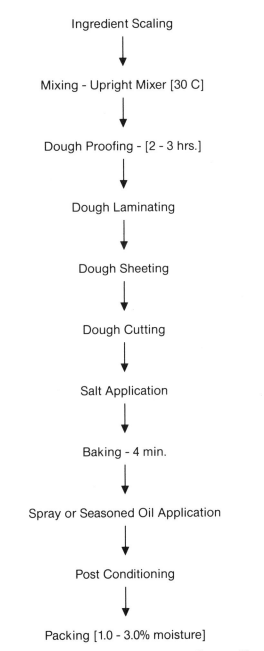

Figure 1-10. Savory cracker production process. Source: Hoseney, 1988.

Typical formulas for various types of crackers (saltine, snack, and grahams) are shown in Table 1-7.

Cookies. In general, cookies are products made from weak soft wheat flours. They are characterized by a formula high in sugar and low in water. Cookie doughs are cohesive but to a large degree lack the extensibility and elasticity characteristic of bread doughs. Relatively high quantities of fat and sugar in the dough allow plasticity and cohesiveness without the formation of a gluten network (Hoseney, 1986). In addition, and again depending on the formulation, cookie dough tends to become larger and wider as it bakes rather than to shrink as does cracker dough.

Perhaps the best way to classify cookies made in North America is by the way the dough is placed on the baking band (Hoseney, 1986).

Rotary molded cookies. For rotary molded cookies, a dry cohesive dough is forced into molds on a rotating roll. As the roll completes a half turn, the dough is extracted from the cavity and falls on the baking oven band (Fig. 1-11A). The consistency of the dough must be such that it feeds uniformly and readily fills all of the crevices of the roll cavity under the pressure exerted in the feeding hopper. At the same time, it must be possible to extract the dough pieces cleanly form the cavity

Table 1-7. Cracker Formulas

Ingredients	Graham	Cheese Cracker Straight Dough	Lunch Biscuit	Saltine Sponge	Saltine Dough	Cheese Snacks Sponge	Cheese Snacks Dough
Flour	80	100	100	65	35	75	25
Graham Flour	20	-	-	-	-	-	-
Shortening	12	12	15	-	-	12	-
Lard	-	-	-	-	11	-	-
Sugar	25	-	3	-	-	-	-
Molasses	5	-	-	-	-	-	-
Invert Syrup	5	-	-	-	-	-	-
Cheese	-	25	-	-	-	-	-
Salt	1	1	1	-	-	-	-
Soda	Variable	0.5	-	-	Variable	-	0.5
Ca Phosphate	-	-	2.5	-	0.5	-	0.25
Ammonium Bicarbonate	0.5	-	-	-	-	-	0.25
Water	20	30	30	29	-	25	5
Sponge	-	10	-	10	-	-	-
Milk Powder	-	-	4	-	-	-	-
Yeast	-	-	-	0.25	-	0.25	-
Paprika	-	-	-	-	-	1	-

Source: B&CMA Handbook, 1981

without their undergoing distortion or forming "tails." Rotary mold dough must be sufficiently cohesive to hold together during baking. Dough spread and rise should usually be minimized. Doughs formulated to meet these requirements are usually fairly high in sugar and shortening but low in moisture. The development of gluten is definitely to be avoided. The typical dough is crumbly, lumpy, and stiff, with virtually no elasticity. Much of the cohesiveness of this type of dough comes from the plastic shortening used (Hoseney, 1986, Hoseney et al., 1988).

Wire-cut cookies. In the production of wire-cut cookies, a relatively soft dough is extruded through an orifice and cut to size, usually by a reciprocating wire (Fig. 1-11B). The dough must be cohesive enough to hold together on the band, yet short enough to separate cleanly when cut by the wire. The rate of extrusion employed is related to the dough's consistency. Here, dough spread during baking is desirable but must be controlled closely. Another type of extruded dough (fruit bars) is similar to wire-cut except that the extrusion of dough is continuous without wires and the orifices are usually designed to produce strips rather than round shapes. Fig bars are made by extruding a fig paste within a tube of dough of

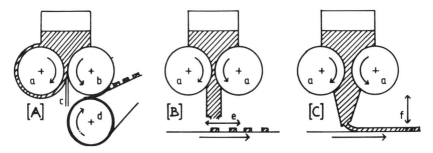

Figure 1-11. Three common methods for cookie production. (A) Rotary method: A dry and crumbly dough is compressed by a grooved roller (a) into biscuit-shaped dies engraved on a forming roller (b). The surplus dough is trimmed from the surface of the (b) roll by means of an oscillating knife (c). The shaped dough pieces are extracted onto a canvas web, which is pressed against the (b) roll by the roller (d). Very little changes in dimension occur during baking. (B) Wire-cut method: A soft dough is extruded through dies by two contra-rotating grooved roller. The extruded dough is cut by a reciprocating wire (a) and usually falls directly onto the oven band. (C) Bar-press method: A soft dough is extruded or co-extruded (fruit bars) and then deposited on a conveyor as a series of parallel ribbons, which are cut to suitable lengths by a cutter (f), either before or after baking. Source: Hoseney, 1988.

Table 1-8. Typical Cookie Formulas

Ingredients (% FWB)	Vanilla Wafer	Chocolate Wafer	Vanilla Sugar Cookies	Oatmeal Cookie	Lemon Crisp	Lemon Snap	Chocolate Snap	Short Bread	Chocolate Chip Cookie	Chocolate Nut Cookie	Ginger Snap	Semi-hard Sweet	Social Tea	Petit Boure
Flour	100	100	100	100	100	100	100	100	100	100	100	100	100	100
Powdered Sugar	70	70	45	70	60	45	35	60	45	55	18	25	29	28
Shortening	30	30	35	35	25	15	-	18	50	50	22	15	15	-
Lard	-	-	-	-	-	-	20	12	-	-	-	-	-	20
Butter	5	8	7	-	10	-	3	-	0.8	-	-	10	15	-
Invert Syrup	10	-	-	-	-	-	3	-	-	-	-	10	-	-
Corn Syrup	-	-	-	-	-	-	6	-	-	-	-	-	-	-
Molasses	-	-	-	-	-	-	-	-	-	-	50	-	-	-
Honey	-	-	-	-	-	-	-	-	-	-	-	4	-	-
Milk Powder	5	6	3	5	2	-	6	1	2.5	2.5	-	-	1.5	-
Egg (Frozen or Powdered)	0.5	2	5	7	4	1.5	-	6	-	10	-	-	1	-
Water	45	-	30	30	35	45	25	15	15	25	5	16	10	0.2
Soda	1	-	1	0.8	0.8	1.2	1.5	2	-	0.5	2	0.75	0.8	-
Calcium Phosphate	0.5	-	0.5	0.25	0.5	-	-	-	0.25	0.25	-	-	-	-
Ammonium Bicarbonate	0.5	-	1	0.5	1	0.2	-	0.5	-	0.25	-	0.6	0.2	0.2
Lecithin	-	-	-	-	-	-	-	-	-	-	-	-	0.2	-
Bisulfite	-	7	-	-	-	-	-	-	-	-	-	0.05	0.1	-
Corn Starch	-	-	-	-	-	-	-	-	-	-	-	5	2.5	-
Potato Flour	-	-	-	-	-	-	-	-	-	-	-	-	-	14
Salt	2	1.5	2.5	1.5	0.8	1	1	1.2	1	1.25	2	1.5	1	0.6
Cocoa Powder	-	10	-	-	-	-	10	-	-	10	-	-	-	-
Chocolate Chip	-	-	-	-	-	-	-	-	50	-	-	-	-	-
Chopped Nuts	-	-	-	-	-	-	-	-	12.5	10	-	-	-	-
Raisin	-	-	-	25	-	-	-	-	-	-	-	-	-	-
Rolled Oats	-	-	-	45	-	-	-	-	-	-	-	-	-	-
Ginger	-	-	-	-	-	-	-	-	-	-	1	-	-	-
Flavor	Butter Vanilla	-	-	-	Lemon Oil	Lemon Oil	Vanilla	Vanilla	Butter Vanilla	Butter Vanilla	Butter Lemon	-	Butter Vanilla	-
Color	-	-	-	-	Yellow	-	-	-	-	-	-	-	Butter	-
Spice	-	-	-	Cinnamon Vanilla	-	-	-	-	-	-	-	-	-	-

Source: B&CMA Handbook, 1981

Table 1-9. Typical Formula for Production of a Co-extruded Fig Bar

Ingredients	Dough	Jam
Flour	100.0	--
Fig	--	100.0
Sugar	40.0	50.0
Invert Syrup	5.0	20.0
Shortening	20.0	--
Eggs	8.0	--
Milk powder	3.0	--
Water	20.0	--
Soda	0.5	--
Calcium phosphate	0.5	0.25
Ammonium bicarbonae	0.5	--
Salt	0.1	0.75
Glucose	--	30.0

Source: B&CMA Handbook, 1981

similar consistency (Fig. 1-11C).

Cutting machine cookies. Another method for cookie production, which is being slowly discontinued from commercial production practice, is the cutting machine. The process and formulation of cutting machine cookies produces the familiar Christmas cookie. A dough with slightly less fat and sugar but more water than rotary mold dough is sent through multiple sheeting rolls and made into a continuous sheet. The sheeting and cutting operations for the short doughs are essentially the same as for cracker doughs. The dough possesses very little elasticity, so shrinkage before cutting is not a problem. Cutting can be similar to the production of crackers in that docking and imprinting of a name or similar pattern is performed before the outline is cut. Typical formulas for the production of various types of cookies are shown in Tables 1-8 and 1-9.

Cakes. Probably the most characteristic and unique product made from soft wheat in North America is cake, traditionally the centerpiece of festive and joyous celebrations (Loving and Brennis, 1981). Cakes, like cookies, are quite high in both sugar and shortening. The major difference between the two is that a cake formula tends to contain much more water (Hoseney, 1986) leading to the production of a liquid batter. An important step in cake production is to incorporate air into the batter as small bubbles. These small bubbles act as nucleation sites for the gas produced during chemical leavening. Double-acting leavening systems are often used in cakes. The first leavening stage reacts during mixing and, as a result, the air cells entrapped during mixing are enlarged. The second leavening acid becomes soluble and, therefore, active at the

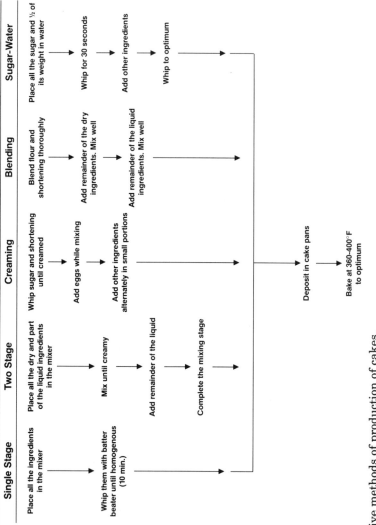

Figure 1-12. Five methods of production of cakes.

baking temperature. Five common methods for cake production are shown in Figure 1-12.

Pie Crusts. Pies are pastries that consist of two distinct components, a relatively thin crust portion which serves to contain the second component, the filling. Pie crusts are low in moisture and high in fat. A good crust is both flaky and tender. The ratio of ingredients, together with the method of preparation prevents formation of a gluten network and results in baked crusts that are more or less friable or flaky (Pyler, 1988).

Hard Pretzels. The hard pretzel is a baked food unique both in its shape and in its hard and shiny outer surface. Pretzels are made from a straight or sponge dough system using a formula which is quite similar to a saltine cracker. Pretzel dough, however, is somewhat drier and, therefore, stiffer. After the doughs are formed, they are first immersed briefly in a hot base (NaOH) bath after which coarse salt is applied. Then they are baked. This unusual treatment, plus baking to a low final moisture content (2–4%), results in the color, sheen, flavor, and texture that is greatly appreciated as a snack and party food in North America (Fig. 1-13). In order for the prod-

Figure 1-13. Pretzels

uct to be acceptable, packaging must maintain moisture content at the low initial level. If it is well-protected, pretzels retain their flavor and crispness for a year or more.

Puff Pastries. Puff pastries are laminated, flaky products with an open structure and a typical flavor of fried goods such as donuts (Pyler, 1988). Puff pastry dough consists of many thin layers of dough separated by fine laminations of fat. A piece of properly made puff pastry dough, 1/4 inch thick, is likely to expand on baking to a height of over 2 inches, an eightfold increase in height. This is an extraordinary amount of expansion, greater than is obtained in the expansion of cake batter to a cake. The gas responsible for the volume increase is steam produced from the dough water. Because the fat layers in the structure are impervious to water vapor, they act as barriers against which the water vapor formed by the underlying dough layer can exert a pressure. In this way, the dough layers expand and separate, giving the lift required. During the latter part of baking, the fat in the layers melts and becomes absorbed by the dough. Because of the light flaky structure which is developed, the baked product has good eating properties.

The lift and the delicate structure of baked puff pastry is a result of lamination by lapping a sheet of puff pastry upon self a number of times. As the number of laps given a dough increases, the number of individual layers increases at a much higher rate. Up to a point, the greater the number of laps (and therefore, number of layers), the greater the lift (McGill, 1975). Lamination beyond that point may decrease the lift and reduce the eating quality of the pastry.

Donuts. Donuts differ from other North American sweet goods in that they are deep fat fried rather than baked (Fig. 1-14). There are two type of donuts: chemically leavened cake and yeast raised. The majority of producers have adopted the use of commercially prepared dry mixes for the production of cake donuts. Mixes avoid the possibility of ingredient scaling errors and require only the addition of liquid during the mixing process to transform them into batters ready for depositing into the frying fat. The base flour for cake donut mixes may be obtained through wheat selection or by blending hard with soft wheat flour to the desired strength and performance level. Flour for cake donut mixes is often dried to aerate and oxidize it so as to help achieve desired performance and stability. The flour is usually unbleached, although light treatment with a maturing agent, such as chlorine dioxide is occasionally employed by some mills. Selection of the chemical leavening system is important. A slow-acting leavening agent is used so that gas evolution is not completed before the second side has expanded in the oil and been cooked or fried to its final shape. Sodium acid pyrophosphate with its controlled reaction rate(s) is usually employed, although other leavening acids such as sodium aluminum phosphate, monocalcium phosphate, dicalcium phosphate, and glucono-delta-lactone are also used in cake donut formulations (Loving and Brennis, 1981, Flick, 1971).

The use of commercially prepared mixes has become as firmly established in the production of yeast-raised donuts as in the case of cake donut production. The extrusion process used to form yeast-raised donuts requires a dough that will extrude readily through the die under air pressure or a vacuum and yet will retain its shape during proofing. This type of dough is most readily obtained by the sponge dough method in which the sponge is fermented at 80°F(27°C) with 6% yeast for 1–2 hours (Pyler, 1988). Proofing conditions should be relatively dry and warm (118–122°F/35–40% RH), to accelerate

Figure 1-14. Donuts

skin formation on the proofing dough pieces. Proof time will normally run from 20 to 25 minutes. A more recent innovation in donut proofing is the application of microwave, or ultra-high frequency energy to heat the product. Microwave energy causes nearly instantaneous, uniform increase in temperature throughout the dough piece rather than the temperature gradient from the surface to the dough interior as is the case in conventional proofing. As a result, normal proofing times of 20–25 minutes are reduced to about 4 minutes (Pyler, 1988). Frying temperatures should be held close to 400°F. At this temperature, frying times range from 45 to 60 seconds per side, depending on donut size.

Refrigerated/Frozen Doughs. By definition, refrigerated dough is an unbaked, flour-based product which must be stored between 32–45°F (Allenson, 1982). As first developed, refrigerated dough was a chemically leavened biscuit with an approximate shelf life of 3 weeks. Today, the refrigerated dough industry encompasses a wide range of products. Such products are packaged in cans or in plastic films and require full baking before consumption (Chen, 1979).

Two types of refrigerated doughs are available. The first is cookie dough packaged in plastic film. It requires only a small amount of leavening action to produce the finished product. Because it has a high solute content (Aw=0.8), it requires little preservation besides refrigeration. The second type is a non-sweet biscuit or bread dough packaged in cans. This type represents the majority of refrigerated dough produced at this time. These are leavened dough products, including a variety of biscuits, cookies, sweet rolls, and dinner rolls. This dough is unique because preservation against both microbial spoilage and loss of leavening action is required.

All refrigerated dough products are leavened chemically. Baking soda ($NaHCO_2$) is the source of CO_2. The leavening acid most commonly used to liberate CO_2 from the soda, sodium acid pyrophosphate (SAPP), is rather unique and particularly useful for refrigerated dough. Its reaction rate is variable (controllable) so one fraction reacts slowly in the dough during mixing and sheeting, preventing the dough from becoming so puffy that it is difficult to fill the cans. A second fraction will react during in-can proofing. Once the can is partially filled with dough and closed, it is warmed at ~90°F for a few hours to trigger the leavening reaction, expand the dough to fill the can and generate a positive can pressure which is essential for preservation of dough quality (Chen, 1979).

Durum Wheat Products

More than 150 types of pasta products are produced and consumed in the United States. These include long goods (spaghetti, linguini, fettucine, etc.), short cut products such as elbows, shells, and noodles, and such specialty shapes as bow ties, rigatoni, lasagna, etc. (Winston, 1971).

The best pasta is produced from semolina. Most U.S. pasta manufacturers prefer semolina that has a fine and uniform particle size rather than a coarse-ground semolina. With fine granulation and narrow size distribution, less trouble is encountered in mixing the semolina and water to form a uniform dough for subsequent extrusion forming. If semolina is not uniform in particle size, the finer particles will tend to absorb water faster than the larger particles, resulting in undesirable white specks in the pasta (Donnelly, 1991).

Pasta Production

Extrusion. Pasta extrusion and drying have evolved to the point where continuous, high throughput presses and dryers are standard throughout the industry. The essential pieces in the process include the continuous press, shaker/spreader, predryer, finish dryer, storage, and packaging.

In the continuous process, water is added to the semolina to give a dough moisture content of approximately 31%. Uniform mixing is carried out in a counterrotating mixing chamber to which a vacuum is applied prior to extrusion. Counter rotating mixing shafts limit balling of the dough, and the vacuum reduces formation of small air bubbles in the dough and limits oxidation of the xanthophyll/lutein pigments. The presence of air bubbles in pasta give the product a chalky appearance and reduces its mechanical strength. Pigment oxidation reduces the attractive yellow appearance of the pasta and its subsequent consumer appeal.

The heart of the continuous process is the extrusion screw, which kneads the dough into a homogenous mass prior to extrusion through a die. Screw speed and temperature control of the dough contributes to the quality of the pasta product. Most modern processes are equipped with sharp-edged screws having a uniform pitch over their entire length. The screw fits within a grooved barrel, which helps the dough move forward and reduces friction between the screw and the inside of the barrel during the extrusion process. Extruder barrels are normally equipped with water-cooled jackets to keep the dough temperature near 40°C during the extrusion process (Donnelly, 1991).

Drying. Moist pasta from the extruder needs to be dried from 31% to approximately 12% moisture so that the product will be hard, retain its shape, and store without spoiling. Regardless of dryer design and temperature-humidity-airflow control, problems can arise if the pasta is not dried carefully. If dried too slowly, the product will tend to spoil or become moldy during drying. On the other hand, if dried too rapidly, moisture gradients will occur, which can cause the product to crack or check. Checking can occur either during the drying cycle or as long as several weeks after the product has been packaged (Hummel, 1966, Donnelly, 1991).

Although microwave technology has been applied successfully to the drying of short goods pasta, microwave drying of long goods pasta is more challenging. The drying system for short goods consists of three stages: a conventional hot air predryer, a microwave-hot air second stage, and an equalized third stage. Advantages claimed for microwave drying include a requirement for 1/3 to 1/4 the floor space required by conventional dryers; reduced drying times; improved product color and cooking quality; reduced plate counts; reduced sanitation costs; and reduced operating costs (Smith, 1979).

Wheat-Based Breakfast Cereals

Breakfast cereal products can be categorized into types based on their use or the physical nature of the product as follows.

Traditional hot cereals that require cooking, are sold in the market as processed raw grains. Wheat grits or rolled oats cereals which require extended cooking and are consumed hot would be examples of this type of cereal.

Instant traditional hot cereals, such as cooked grains, require only boiling water to complete their preparation. Wheat and oat based cereals are, again, examples.

Ready-to-eat cereals require no cooking prior to consumption. They are manufactured from grain products having been cooked and modified such that they may be subdivided into flaked, puffed, or shredded products.

Ready-to-eat cereal mixes are cereals combined with other grains, legumes or oil seeds and dried fruit products. Granola cereal mixes best describe this category.

Miscellaneous cereal products include cereal products

which cannot be included in any of the above types because of specialized production processes or end uses (e.g. cereal nuggets and baby foods) (Tribelhorn, 1991).

Most breakfast cereal products contain high proportions of cereal grains and have very few additives. When present, additives are primarily used to improve the texture or to change functional characteristics of the final product. Because these types of cereal products are often the only food consumed for the morning meal, they are supplemented with vitamins and minerals to improve their nutritional quality. Addition of these ingredients is based on a specific percentage of daily required intake for adults.

Wheat-Based Products Requiring Cooking. Wheat-based cereal products are wheat middlings (farina) obtained from the roller milling process. They are endosperm pieces, which ideally are free from bran and germ. Careful attention is given to the particle size range of these middlings. The most acceptable products have the following minimum particle size specifications.

Through U.S. #20 – 100%
Through U.S. #45 – <10%
Through U.S. #100 – <3%

Hard wheats have been found acceptable for making farina-based cereals because the end product does not get pasty during cooking. One disadvantage to hard wheat is the longer process time required to achieve a good cook, i.e., to gelatinize the starch in the wheat. Generally, no other cereal or cereal fractions are used as ingredients in these formulations (Tribelhorn, 1991).

"Quick cook" cereals are prepared by steaming grain particles at elevated temperatures and pressure followed by flaking. Generally, this method of processing reduces the preparation time of the cereal to about one-third the time of traditional (raw) cereals. Particle size during steam cooking is critical to the creation of an acceptable product because the extent of gelatinization during processing and, consequently, the mouthfeel are affected by particle size (Matz, 1959). The smaller the particle, the more surface area available for heat transfer and, thus, the better degree of cook attainable.

RTE Cereal Grains: Flakes. Flaked cereals are manufactured mainly from corn and wheat. A typical flaked cereal might have as a formulation approximately 90% cereal components plus 8% sugar, 1% salt, and 1% malt. Recently developed flaked breakfast cereals have included multiple cereal formulations and cereal formulations in combination with other seeds such as soybean. Flaked cereal grains were the first form of ready-to-eat cereal products available to the consumer. Processes to make the flakes are simple and result in products that are both well cooked and have acceptable flavor. The traditional method of manufacturing a flaked corn product begins by cleaning the grain, followed by milling to break the whole grain into pieces that are one-third to one-half the size of the original whole kernel. These pieces are mixed with other ingredients as required and steamed under pressure for 2 hours or longer. The steamed mass is then broken into its original sizes and partially dried under carefully controlled conditions. The resulting cooked and dried pieces are then either tempered for 24 hours or directly flaked between steel rolls (Harper, 1981). The resulting flakes are then dried and toasted at high temperature to give a suitable flavor and color.

Variations on this process have been used to improve upon and introduce new flaked cereal products. The primary difference between these processes and the traditional flaking process is the replacement of the steam-cooking step with extrusion processing. In addition to cooking the grains, extrusion allows formation of uniform pellets and, thus, flakes which can be made from single or multicomponent formulations. The typical process flow to make flaked products are shown in Figure 1-15.

RTE Cereal Grains: Puffed Products. Certain intact grains (wheat and rice) and a variety of cereal-based doughs expand to several times their original volumes when water within them undergoes a rapid transition from the liquid to vapor (steam) state. This forms the basis for producing puffed cereals. Three different processes are currently in use. Regardless of the method of production, puffed cereals possess a cellular structure with a uniformly delicate and brittle texture.

Gun puffing, the process used to make puffed wheat, heats the starting material under pressure to temperatures well above 100°C. Instantaneous release of this pressure causes the super-heated moisture within the kernels to flash off as steam, gelatinizing the starch, expanding, drying and setting the product. Oven puffing subjects the starting kernel or dough pieces to intense heat at atmospheric pressure. Vaporization of the water in the product accomplishes essentially the same results as in gun puffing but with slightly less relative expansion. HTST extrusion cooking, the newest technology applied to the task, applies heat, shear and pressure to the dough in process before it is forced out through a forming die to atmospheric temperature and pressure. The immediate vaporization of product moisture expands and sets the product (Tribelhorn, 1991). Of all the technologies, extrusion cooking is by far the most versatile in terms of the range of possible product sizes, shapes and formulations.

RTE Cereal Grains: Shredded. Shredded breakfast cereals are made from whole cereal grains, primarily wheat. However, new shredded products are being developed which

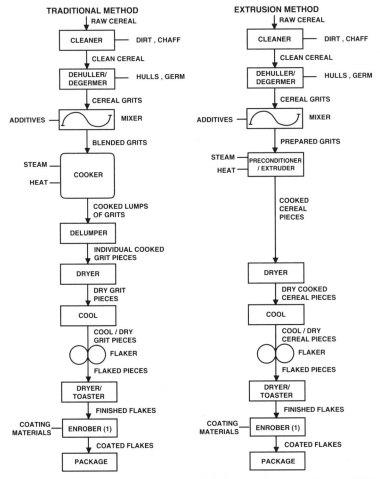

Figure 1-15. Process lines for cereal flake production. Source: Tribelhorn, 1991.

use alternative cereals, such as rice or corn and some are now offered with soft center, fruit-flavored confections.

Cleaned grain is boiled in water at atmospheric pressure until tender. The moisture in these cooked grains is allowed to equilibrate in a tempering process for several hours before processing the grains into shreds with specially designed, paired steel rolls running with no gap between them. For this step, one roller is smooth and the other roller is corrugated radically to cause the formation of the small strands. The strands exiting the rollers are layered then cut into the full- or bite-sized (dimension) pillow-shaped products and toasted or baked. The finished shaped products are then packaged or coated as required. Extrusion is also being used to make shredded products (Matz, 1959, Fast, 1987).

References

Allenson, A., 1982, Refrigerated Doughs, Bakers Digest, 56(5):22-24.

Anonymous, 1992, Six to Eleven: 70th Anniversary Commemorative Issue of Milling and Baking News, Sosland Publishing Company, Kansas City, MO.

Banasik, O.J., 1981, Pasta Processing, Cereal Foods World, 26(4):166.

Biscuit and Cracker Handbook, 1981, Biscuit and Cracker Manufacturers Association, Washington, DC.

Chen, R. W., 1979, Refrigerated Doughs, Cereal Foods World, 24(2):46-47.

Donnelly, B.J., 1991, Raw Materials and Processing, in Handbook of Cereal Science and Technology, K. Lorenz and K. Kulp ed., Marcel Dekker Inc., NY, NY, p. 763-792.

Faridi, H., 1991, Soft Wheat Products, in Handbook of Cereal Science and Technology, K. Lorenz and K. Kulp ed., Marcel Dekker Inc., NY, NY, p. 683-740.

Fast, R.B., 1987, Breakfast Cereals: Processed Grains for Human Consumption, Cereal Foods World, 32, (3):241-244.

Federal Register, June 30, 1987, Volume 52, No. 125, pp.24430-24431.

Flick, O.B., 1971, Production of Yeast-Raised Doughnuts, Proceedings of the American Association of Bakery Engineers:174-183.

Harper, J.M., 1981, Extrusion of Foods, Volume II, CRC Press, Inc., Boca Raton, FL, pp. 78-86.

Hoseney, R.C., 1986, Principles of Cereal Science and Technology, American Association of Cereal Chemists, St. Paul, MN, pp. 245-276.

Hoseney, R.C., Wade, P., and Finley, J.W., 1988, Soft Wheat Products, in Wheat Chemistry and Technology, 3rd ed., Y. Pomeranz (Ed.), American Association of Cereal Chemists, St. Paul, MN, pp. 407-456.

Hummel, C., 1966, Macaroni Products: Manufacture, Processing, and Packing, 2nd ed., Food Trade Press, Ltd., London, p. 1.

Kulp, K., 1991, Breads and Yeast Leavened Bakery Foods, in Handbook of Cereal Science and Technology, K. Lorenz and K. Kulp ed., Marcel Dekker Inc., NY, NY, p. 639-682.

Kulp, K., 1983, Technology of Brew Systems in Bread Production, Bakers Digest, 57(3):20, 22-23.

Kulp, K., and Dubois, D.K., 1982, Breads and Sweet Good in the United States, American Institute of Baking Research Department Technical Bulletin, 4(6).

Loving, H.J., and Brennis, L.J., 1981, Soft Wheat Uses in the United States, in Soft Wheat: Production, Breeding, Milling and Uses, W. T. Yamazaki and C.T. Greenwood (Eds.), American Association of Cereal Chemists, St. Paul, MN, pp. 169-207.

Manley, D.J.R., 1983, Technology of Biscuits, Crackers and Cookies, Ellis Horwood Ltd. Publishing Co., Chichester, UK.

Mattern, P.J., 1991, Wheat, in Handbook of Cereal Science and Technology, K. Lorenz and K. Kulp ed., Marcel Dekker Inc., NY, NY, p. 1-54.

Matz, S.A., and Matz, T.D., 1978, Cookie and Cracker Technology, AVI, Westport, CT.

Matz, S.A., 1959, The Chemistry and Technology of Cereals As Food and Feed, AVI Publishing Company, Inc. Westport, pp. 547-566.

McGill, E.A., 1975, Puff Pastry Production, Bakers Digest, 49(1), 28-28.

Miller, W. Kirk, June 30, 1987, Grain Standards; Official U.S. Standards, Handling Practices and Insect Infestation; Final Rules, Federal Register, 52(125):24414-24442. Federal Grain Inspection Service, USDA.

Ponte, J.G., Jr., 1981, Production Technology of in Variety Breads in United States, Miller, B.S. (Ed), American Association of Cereal Chemists, St. Paul, MN, pp. 9-26.

Pyler, E.J., 1988, Baking Science and Technology, Siebel Publishing Co., Chicago, IL.

Redfern, S., Brachfeld, B.A., and Maselli, J.A., 1964, Continuous Mix Temperatures, Cereal Science Today, 9:190-191.

Semling, H.B., Jr., 1988, Continued Growth in 1988, Baking Industry, 101(1):21-23.

Smith, F.J., 1979, Microwave-Hot Air Drying of Pasta, Onions and Bacon, Microwave Newsletter, 12 (6).

The Canadian Wheat Board, 1990-1991, Annual Report, 1992, Winnipeg, Manitoba.

Tribelhorn, R.E., 1991, Breakfast Cereals, in Handbook of Cereal Science and Technology, K. Lorenz and K. Kulp ed., Marcel Dekker Inc., NY, NY, p. 741-762.

U.S. Department of Agriculture, May 1, 1988, FGIS Grain Inspection Handbook, Book II, Wheat.

USDA, September, 1957, Grain Grading Primer, Miscellaneous Publication No. 740.

Winston, J., 1971, Macaroni, Noodles, Pasta Products, in Publishing Corporation, New York.

Zelch, R., 1988, American Institute of Baking, Manhattan, KS, Personal Communication.

CHAPTER 2

Wheat Usage in Mexico and Central America

Roberto J. Peña

Introduction

Wheat was brought to America in the 16th century by the Spaniards to what would become Mexico in 1530 and to the future United States in 1585 (Pomeranz, 1987). As the Spaniards explored the land south of Mexico, they left behind the settlers who introduced wheat cultivation and consumption (in the form of bread) to the Central American region (i.e., south of Mexico and north of Colombia).

Wheat rapidly established itself as a commercial crop in North America and an important source of nutrients for the population. This was not the case in Central America (CA), mainly because the cultivation of other crops better adapted to the warm, tropical climate of the region than wheat has been economically more advantageous. Additionally, wheat has not been as important as maize or rice to the population of this region as a basic food. Wheat is cultivated in Mexico but only in a small area of Guatemala. There has been recent interest in wheat cultivation in El Salvador (J. A. Orellana, *personal comm.*) and in Costa Rica (R. Alfaro-Monge, *personal comm.*). Because Mexico is much larger than the entire CA region in population, food utilization of wheat and wheat processing, it will be treated separately in dealing with various aspects of wheat usage.

Wheat and Flour Usage

MEXICO

Wheat is a staple food for the Mexican population. It is consumed mainly in the form of bread products which vary in their type across the different regions of the country. The existence of a variety of wheat-based products, white breads and sweet breads, resulted from the combination of two main factors: the fusion of races, food, and cultural habits between natives (Mayan, Aztec) and European immigrants (Spanish, French, Italian), and the widespread establishment of wheat, both as a subsistence crop for low income farmers in rural areas and as a large scale commercial crop linked to urban and industrial development.

Wheat production in Mexico fluctuates around 4.0 million tons, approximately 25% of the amount of maize, the main staple food (Table 2-1). Mexico produces predominantly hard to semi-hard wheats, with lesser amounts of soft and durum wheats. More than 90% of the wheat harvest occurs during spring and fall. It is purchased annually by the mills directly from the farmers, at a minimum subsidized price fixed by the government (Mielke 1990). Price premiums and penalties are applied based on grading factors such as test weight, grain moisture, grain soundness, and cleanliness.

Hard wheat imports represent between 5 and 15% of the

Table 2-1. Wheat and Maize Production, Imports and Food Supply, 1990.

Country/Region	Production		Imports		Supply	
	Wheat	Maize	Wheat[a]	Maize	Wheat	Maize
	----------------------------(1000tons)----------------------------					
Mexico	3931	14635	420	4124	3545	11400
C A[b]:	33	2973	722	573	765	2435
Belize	0	18	12	0	11	5
Costa Rica	0	72	130	205	121	59
El Salvador	0	603	106	65	105	488
Guatemala	32	1293	219	130	233	1072
Honduras	1	559	98	75	113	498
Nicaragua	0	214	73	55	98	253
Panama	0	214	84	43	84	60
Total[b]	3964	17608	1142	4697	4310	13833

[a] Includes wheat flour
[b] Calculated

Source: FAO AGROSTAT 1992
Wheat Import-supply figures do not match due to stocks from previous year.

total wheat used for food. Mexico is the only country in the Mexico and CA region where wheat is used as feed grain. Approximately 600,000 tons of locally produced wheat went to the feed industry in 1990 (FAO 1992). The total food use of wheat in Mexico is several times that of the entire CA region (Table 2-1), due mainly to differences in population. This is so in spite of the fact that per capita consumption in Mexico is rather low if compared to that of maize (Table 2-2).

CENTRAL AMERICA

Depending on the country, maize, wheat, rice and grain sorghum are, to different extents, important staple foods for the population. Hard wheat comprises the majority of the wheat consumed in this region. The use of soft and durum wheat is low and limited in most regions. Wheat usage varies both with the country as well as with the distribution of the population between rural and urban areas. In general, wheat is used for the manufacture of white, crusty type breads and sweet breads. Lesser amounts of pan-type bread, soft buns, cookies, pastry, cake, and pasta products are also consumed, particularly in urban areas.

Most of the baking industry is located in urban areas and consists of many family-type to medium-sized bakeries, and of some few well mechanized bread, cookie and pasta making plants. In the small bakeries most dough making operations are carried out by hand. In medium-sized bakeries, dough mixers and ovens are usually the only equipment used.

In Central America, most of the area sown to cereals is as-

Table 2-2. 1990 Per Capita Consumption of Wheat, Maize and Rice

Country/Region	Per capita consumption		
	Wheat	Maize (Kg/year)	Rice[a]
Mexico	40.0	128.7	7.2
C A:			
Belize	57.3	13.7	25.9
Costa Rica	40.2	19.6	55.2
El Salvador	19.9	92.9	10.0
Guatemala	25.3	116.5	5.4
Honduras	22.0	97.0	9.8
Nicaragua	25.4	65.3	36.1
Panama	34.5	24.8	53.2

[a]: Husked equivalents

Source: FAO AGROSTAT, 1992

signed to maize (FAO 1992). Wheat is cultivated only in Guatemala (Table 2-1). Consequently all countries in CA import wheat and maize (chiefly from the USA); wheat for food consumption and maize for both food and feed. Guatemala imports more than 75% of its food demand of wheat, while the rest of the CA countries import all of their wheat needs (Table 2-1).

The per capita consumption of wheat is low in all countries of CA, ranging from 20 to 57 Kg/year (Table 2-2). It is smaller than that of maize in El Salvador, Guatemala, Honduras and Nicaragua, but larger in Belize, Costa Rica and Panama. Rice is more important as a staple food than wheat in Costa Rica and Panama (Table 2-2).

Milling Industry

MEXICO

The large difference in the food supply of wheat between Mexico and CA countries is reflected in the relative number of flour mills (Table 2-3). In Mexico, the wheat milling industry consists of 133 mills, with an installed capacity of 4.9 million tons of flour per year (F'uguemann 1991). This is sufficient to mill all the wheat produced locally. The industry is integrated into seven regional organizations distributed throughout the country, and in turn within the National Association of Manufacturers of Flours (F'uguemann 1991). Approximately 30–40% of the mills are small, producing less than 100 tons of flour per day, while 20–30% are large mills, with flour production of ≥ 300 tons per day (Barranco-Fernandez 1991, McWard 1991).

BELIZE

Before the construction of the only mill in Belize (1976), all wheat flour was imported from the USA and Canada. Maple Leaf Mills International Ltd. purchased this locally owned mill in 1985, revitalized it, and added a new compact pneumatic milling unit in 1988. The mill utilizes both hard red winter and dark northern spring wheats imported from the USA Hard wheat flour (patent, all purpose, and baker's) from hard wheat is sold in large bags to bakeries, or in small units for domestic use. (Anonymous 1990a).

COSTA RICA

Annual wheat imports by Costa Rica range from 130,000 to 140,000 tons. Part of the wheat imports are through PL480 but the majority is purchased. Imports include hard, soft and

Table 2-3. Wheat Mills in Mexico and Central America (CA)

Country/Region	No. of Mills
Mexico	133
C A:	36
Belize	1
Costa Rica	2
El Salvador	2
Guatemala	24
Honduras	2
Nicaragua	2
Panama	3

durum wheats (C. Salas-Fonseca, *personal comm.*). The two mills in Costa Rica; Molinos de Costa Rica at Alajuela and Fabrica de harinas de Centro America at Puntarenas, process 81% and 19% of the wheat supply, respectively (Anonymous 1989). These two mills have sufficient capacity to satisfy the present wheat flour demand.

The Costa Rican government controls the price of wheat and wheat by-products, including flour. The National Production Council (Consejo Nacional de Producción, CNP) sells wheat directly to mills. The resulting flour is priced in four categories: bread, cookie, pasta and whole wheat flour. Flour and bread prices are regulated based on the price of the wheat imported into the country (Anonymous 1988).

EL SALVADOR

El Salvador imports both hard red spring and soft red winter wheats from the USA, mainly through the PL480 credit plan. Small donations of wheat are received from Canada, the EC and Japan (Anonymous 1991c). El Salvador currently has two wheat flour mills in operation, both in San Salvador: Molinos de El Salvador (MOLSA), with a milling capacity of 340 tons per day, is partially owned by The Pillsbury Co.; Fabrica Molinera Salvadoreña (FAMOSSA) produces about 380 tons per day. Total annual wheat flour production is 120,000 tons (Anonymous 1991c).

GUATEMALA

This country produces only soft wheat in amounts accounting for 25% of the total wheat demand. Guatemala imports approximately 158,500 tons of hard wheat annually from the USA through PL480 and GM102 programs (Anonymous 1990b). There are 24 mills in operation with a total capacity of approximately 1,760 tons per day and an average capacity per mill of 80 tons per day. Local mills must purchase their quota of local wheat before they are allowed to import (Anonymous 1990b). Hard wheat is milled directly into flour or blended with small amounts of locally produced soft wheat. Thus, hard wheat flour predominates in the baking industry, although semi-hard and soft type flours are also available. Guatemala imports roughly 22,500 tons of durum wheat for the manufacture of pasta products (H. E. Carranza, R. Mora and V. Azañon, *personal comm.*).

HONDURAS

Honduras imports roughly equivalent amounts of hard and soft wheat. The two mills in the country supply the flour demands of the baking industry (M. López, *personal comm.*). Approximately 63,000 and 53,000 tons of soft and hard wheat flour, respectively, are consumed annually (Ministerio de Economía de Honduras 1992).

NICARAGUA

Hard wheat is purchased through the PL480 program. Two mills supply all of the country's wheat flour needs (Agurto-Vilchez and Utting 1985).

PANAMA

The bulk of the wheat imported comes from the USA. There are three mills in the country, with a fourth under construction (U.S. Wheat Associates 1992). The largest (Harinera Panamá) has a milling capacity of 200 tons/day. With a capacity of 120 tons/day, General Mills de Panama (the second largest) is undergoing expansion. The milling industry produces six types of hard wheat flour, including strong and medium-strong types for the manufacture of bread. Small amounts of soft flour are milled for the production of cookies, pastry and cakes. Panama does not import durum wheat. Apparently pasta is manufactured with hard wheat farina.

Baking Industry

MEXICO

The baking industry in Mexico is very diverse, ranging from small family bakeries to large, modern baking plants. Most of the small bakeries are located in small villages and towns of the rural area. They use whole wheat meal and flour for the labor-intensive preparation of breads, mainly sweet-type breads with a dense crumb (Fig. 2-1). In contrast to white breads, which stale rapidly, these breads can be stored for several days with minimal staling due to the high proportion of sugar in their formulas. In the urban areas, small to large semi-mechanized bakeries manufacture a variety of baked products: soft buns, hard rolls, whole meal breads, a large number of international white breads, a variety of sweet goods (including sweet breads lighter in crumb than those made in rural areas), cookies, pastries and cakes. In the urban areas baked products are purchased daily, mainly from retail bakeries. In Mexico there are several modern mechanized baking plants manufacturing a large variety of breads, cookies and crackers. Small (≈5,000 tons/year) to large (≈60,000 tons/year) pasta plants produce a wide variety of pasta and noodles (C. A. Monroy, *personal comm.*).

Until now, end use quality has not been a major factor considered in the development of wheat cultivars. Consequently, the farmers encourage wheat research institutions to develop high-yielding cultivars. Therefore, the production of high-yielding wheats of low protein and inferior quality has, traditionally, predominated in Mexico. Recently, as the North America Free Trade Agreement (NAFTA) is being enacted and

Figure 2-1. Sweet breads made by rural populations of Central Mexico.

a substantial increase in the imports of quality wheat to satisfy the industry's quality needs is foreseen, the improvement of quality of Mexican wheats has become a major research concern.

The baking industry has established five wheat end-use quality groups based on hardness class and gluten strength "type" (Table 2-4). Gluten strength "type" is determined with the alveograph (gluten strength value W and tenacity/extensibility ratio, P/G). Group 4 wheats tend to disappear from local cultivation because their quality (tenacious gluten type) is undesirable.

The most common breads made with hard wheat flour are: *pan blanco* (white bread), a term covering a variety of breads with medium dense crumb and hard to soft crusts; and *pan dulce* (sweet bread), including sweet breads made with yeast-leavened doughs.

BELIZE

The baking industry consists of many small bakeries which sell directly or through retail stores and supermarkets. There are no wholesale baking companies in Belize (Anonymous

Table 2-4. Wheat Quality Groups in Mexico and 1992 Usage Distribution

Quality Group	Grain Type	Gluten Strength type	End-Use	End-Use Distribution (1000 tons)	%
Group 1	Hard	Strong, balanced to extensible (B--E)	Pan-type bread; to correct weaker wheats	615	15
Group 2	Hard to semi-hard	Medium-strong, B-E	French and Spanish type breads (e.g. *bolillo, telera*); light crumb & flake dough sweet breads	1,845	45
Group 3	Soft	Weak, extensible	Cookies; cakes; tortilla; sweet breads	1,025	25
Group 4	Hard to semi-hard	Strong to weak, tenacious	Cakes; dense-crumb breads and pastry	410	10
Group 5	Durum	Medium strong to strong	Long and short Italian pasta; noodles	205	5

Source: Department of Agropecuary Politics of the Secretary of Agriculture and Hydraulic Resources, Mexico.

1990a). Soft wheat-based products such as cookies, crackers, cakes, among others, are imported primarily to satisfy the needs of the tourism industry (Anonymous 1990a). Durum wheat is not milled locally. However, durum wheat-based products such as pasta and noodles, among others, are imported, again primarily to satisfy tourism industry needs (Anonymous 1990a).

COSTA RICA

Family-type and small to medium sized semi-mechanized bakeries comprise most of the baking industry of Costa Rica. French-type bread, known as *bollo*, consumed mainly with breakfast, as well as sweet bread called *pan dulce*, are the most popular baked products. However, the production of pan-type bread is on the rise (Perts et al. 1981).

In 1992, Costa Rica imported 105,000 tons of hard wheat from the USA for the manufacture of pan-type bread, hard rolls (common bread), whole meal bread, French-type bread and flour tortilla; 17,500 tons of soft wheat for sweet bread, cookies, crackers and cakes; and 15,600 tons of durum wheat for the manufacture of long (spaghetti and macaroni) and short (noodles, small shells, etc) pasta products (F. Trejo-Ballesteros and C. Salas-Fonseca, *personal comm.*)

EL SALVADOR

The breadmaking industry of El Salvador comprises many small bakeries, medium-size mechanized bakeries and bakeries that combine hand-manufacture and mechanized breadmaking. Flour milled from imported hard red spring wheat is used to produce pan-type bread, hot dog and hamburger buns, French-type and variety breads. Soft wheat flour is used for the preparation of sweet breads, cookies and cakes, as well as blended with hard wheat flour (30–40% soft: 70–60% hard) for the preparation of certain types of bread (Anonymous 1990c).

GUATEMALA

About 65% of the Guatemalan baking industry consists of small, family-owned bakeries, with a monthly average usage of 7 tons of flour. In these bakeries, all processing is by hand. All of the ingredients are weighed together and hand mixed and kneaded into doughs. Wood-fired clay ovens are common in these small bakeries. Approximately 25% of the baking industry consists of small bakeries combining hand-manufacture with semi-mechanized breadmaking, (i.e., using at least dough mixers). A limited variety of breads are produced in these bakeries. The two main types, both made with yeast-fermented doughs, are sweet bread, known as *pan de manteca* (lard bread) or *pan dulce,* and white bread, known as *pirujos* (white bread with soft crust and dense crumb). *Pan dulce* constitutes 40% of the bread consumed in the country, while the remainder is white bread (Anonymous 1990b). In small bakeries, bread is sold at the bakery shop with no commercial distribution, generally from a previous purchase order placed by the client. In turn, the client distributes baked products door to door, or in small food stores (Anonymous 1990b; Anonymous 1991a). Large breadmaking plants use modern baking equipment to produce *pan blanco, pan desabrido* or *pan salado* (all referring to French-type bread), *pan de sandwich* (pan-type bread), and hamburger buns (Anonymous 1991a). Pasta plants process durum wheat semolina into the Italian-type long and short pasta products which are consumed in urban areas (H. E. Carranza, R. Mora and V. Azañon, *personal comms.*).

HONDURAS

Wheat is consumed in the urban areas of Honduras mainly, where the bulk of the baking industry is located (M. López, *personal comm.*). Soft wheat flour is used for the manufacture of cookies, cakes called *queques* (layer cakes mainly) and sweet breads. Yeast-leavened and unleavened sweet breads of various types and shapes are very popular for breakfast, particularly *semita*, a sweet dense yeast-leavened bread. A small number (2–4) of cookie-making plants manufacture sugar cookies (wire-cut or rotary-molded) and soda crackers. Cookies are also imported. The per capita consumption of wheat-based cookies is not large, because maize-based cookies are more popular (Futrell and Jones 1982). Soft wheat flour is both packed and sold in stores for production of the above products as well as for the preparation of flour tortilla and of *queques* to celebrate special occasions. (M. López, *personal comm.*). White bread, called *pan blanco* or *bululo,* is the most popular bread in Honduras. French-type bread and pan-type bread is occasionally consumed, primarily in the urban regions. Bread, preferably sweet bread, is consumed mainly with breakfast (M. López, *personal comm.*).

There are one or two small pasta plants that produce long and short pasta using hard wheat farina. Durum wheat-based pasta products (long, short pasta, Italian noodles) are imported (M. López, *personal comm*). The consumption of pasta products, locally manufactured and imported, is mainly associated with restaurants. Because the country has a significant proportion of oriental immigrants, oriental-type noodles (imported) are popular in urban areas (M. López, *personal comm.*).

NICARAGUA

For many years wheat bread was an urban food. Managua, the capital city, contained western-type supermarkets where many wheat-based foods could be purchased only by the population with high income (Barraclough 1984). In recent years, consumption of wheat products has increased to the point that wheat is now considered a basic food item. The most popular wheat-based products are French-type bread known as *bollo* (consumed mainly with breakfast) and the sweet bread called *pan dulce*. The consumption of pan-type bread has increased significantly in recent years (Perts et al. 1981). In most cases, bread is not sold at the bakery. It is sold at grocery stores called *pulperias* or distributed by door to door sellers who pick up the bread at the bakery, (Agurto-Vilchez and Utting 1985).

PANAMA

Because of the large number of North Americans residing in Panama permanently or for long periods of time, the Panamanian bakery industry produces many U.S.-type products (U.S. Wheat Associates 1992).

The breadmaking industry of Panama consists of a large number of small bakeries and a few breadmaking plants. There are roughly 400 bakeries in Panama, with 225 located within Panama City and its suburbs. In general, bakers start as apprentices, loading ingredients and moving doughs to work tables. They learn dough handling operations until they become *panaderos* (white bread bakers) or *dulceros* (sweet bread and cookie bakers). *Panaderos* and *dulceros* manufacture baked products by hand, using little equipment except for the dough mixers (Fig. 2-2).

There are eight large breadmaking plants in the country, the largest one with a usage of 11 tons of flour per day. It produces bread for fast-food outlets and restaurants. Hamburger and hot dog buns are made using straight-dough methods and short fermentation times. Pan-type bread is produced using the sponge and dough method (U.S. Wheat Associates 1992).

White bread is produced in small bakeries using the straight dough method and long fermentation times (from 6 to 16 hr depending on the type of bread). Long fermentation times are required due to the common use of low levels (<2 %) of fresh yeast. The more popular white breads are: the French-type bread known commonly as *flauta*, long flute-shaped loaves, and *michitas*, shaped as the latter but smaller. Hard-crusted breads of various sizes and shapes are popular also (Fig. 2-3).

Of the large variety of sweet breads manufactured in the country, *rosquitas*, a hard ring-shaped sweet bread, is very popular. *Rosca de huevo*, a large, soft, ring-shaped, medium-dense sweet bread enriched with eggs, is traditional for special occasions such as Christmas, New Year's Eve and Mothers' Day (U.S. Wheat Associates 1992). Soft wheat products (cookies, cakes and pastries) are consumed mainly in urban areas. Annual per capita consumption is low, ≈5 kg. Pasta consumption is low, probably less than 5 kg/capita/year.

Products Made from Hard Wheat

The most popular white breads and sweet breads in Mexico and Central America are made with hard wheat flour or hard and soft wheat blends. White breads are manufactured by

Figure 2-2. Top view of the baking shop of a medium-size bakery of Panama (Courtesy: El Mundo del Pan, Mexico).

straight dough methods using formulas typical of medium-dense USA breads. They may have a hard or soft crust. They are generally consumed the day of purchase, because these products stale rapidly. Size, shape and specific volume vary with location, food habits of the population, and the availability of appropriate types of wheat and the type of bakery. Sweet breads are more hand-manufactured baked products. They are generally prepared by straight dough methods. Ingredients, fermentation time, size, shape, and decoration in many cases make the product typical of a specific region.

White Breads

Pan blanco is the generic name given in Mexico to popular white bread, made with a relatively lean formula and low level of yeast, using the straight dough procedure. Hard wheat flour (100%), fresh yeast (2%), salt (1–2%) and water ($\approx$60%) are mixed and the resulting dough fermented for short periods of time (20–30 min) in two to three stages: after mixing, after dividing and punching, and after dough shaping. In general, regardless of size and shape, the bread is light with an open and nonuniform crumb. The bread's flavor and texture are

Figure 2-3. A sample of French-type bread varieties of Panama (Courtesy: El Mundo del Pan, Mexico).

best when very fresh. Therefore, it is purchased daily, mainly in retail bakeries. The most popular varieties of *pan blanco* are *bolillo* and *telera* (Fig. 2-4).

Tortilla de harina (flour tortilla) is a single-layered, unleavened bread eaten rolled and filled with vegetables and meats. It is consumed daily by the urban and rural inhabitants of northern Mexico. Its consumption is spreading, becoming popular in central Mexico, mainly as an urban food. A high-quality flour tortilla should be flexible to allow rolling without cracking. Tearing quality is important and reflects chewability. Home-made flour tortilla is commonly consumed the same day it is made because of its short shelf-life. Commercially produced flour tortillas keep well for a week under refrigeration. Shelf-life can be improved by the addition of emulsifiers to the formula.

In Mexico, flour tortilla is prepared mainly with hard wheat flour. However, soft wheat flour can be suitable, particularly when the product is made at home. In industrial production, hard wheat flour doughs have better machinability than do soft wheat flour doughs (Qarooni et al. 1993, Serna-Saldivar et al. 1988).

The formula for flour tortilla is simple, consisting of wheat

Figure 2-4. Popular white breads of Mexico. Bulk presentation of *telera* in a self-service bakery.

flour (100%), shortening (12%), salt (1.5–2%) and water (≈40%). Its preparation entails mixing all the ingredients into a dough and dividing the dough into 20- to 30-g pieces. These are allowed to relax for 10 to 15 min (to facilitate subsequent processing), then rolled or pressed into disks 12–15 cm in diameter. Because both thick and thin versions of this product are consumed, thickness ranges from 0.2 to 0.5 cm. Baking takes place on a hot griddle or hot plate at approximately 200°C. When the tortilla puffs (after 15–20 sec, depending on its thickness) it is turned to bake the other side for 10–15 sec.

Flauta and *michita* are the most popular varieties of white bread consumed in Panama. These, made using the straight-dough method, are similar in shape to baguettes (Fig. 2-5) (U.S. Wheat Associates 1992). Formulas are similar to lean French breads: hard wheat flour (100%), shortening (1.5%), salt (2%), sugar (1%), yeast (1%) and water (≈54%).

Sweet Breads

Sweet breads are very popular in Mexico. They are generally made with yeast-leavened doughs, rich in shortening and sugar. Among the sweet breads, the most popular type is

Figure 2-5. *Michita*, popular French-type bread of Panama (Courtesy: El Mundo del Pan, Mexico).

known as *bizcocho*, a medium-dense, soft and spongy sweet
bread derived from the French *bisscotte*. There are many varieties of *bizcocho*, some of which are characteristic of specific
geographic regions. The formula for *bizcocho* is in practice the
same for all varieties. The difference among types is based
mainly on the shape and decoration given to the product.

The basic *bizcocho* formula consists of medium-strength
hard wheat flour (100%), sugar (20%), shortening (30%), salt
(1.2%), whole eggs (6%), dry yeast (1.5%) or fresh yeast (5%),
water or milk (35–40%) and artificial flavor and color, as desired.

After using a dough mixer to incorporate the eggs, salt, color
and flavor ingredients, yeast (if fresh), half of the sugar and
water or milk, flour and yeast (if dry) are added, followed by
shortening and the rest of sugar. Mixing is for 10 min to incorporate fat and sugar, and continues until a soft, cohesive
dough has been formed. The dough is allowed to ferment for 30
min, divided, and shaped then proofed for 40 to 50 min. at
30°C. Prior to baking (15 min at 180–200°C), dough pieces are
decorated according to variety. Decoration consists of a top
layer of a decorating paste or spreading the top with raisins,
sesame seed or ground dry molasses. A paste, made with sugar
glass, shortening and flour, softened with whole egg, is commonly used to decorate the surface of some varieties.

Popular *bizcocho* varieties include: *mosca*, a raisin-topped
loaf; *concha*, bread with a layer of paste resembling a shell;
and *amapola*, resembling a flower (Fig. 2-6). *Bizcocho* dough is
also used to make *brioche*; this variety is sweeter than the
French brioche (Anonymous 1991b).

Semita, a popular sweet bread in El Salvador, with a round
shape and dense crumb, is made with two layers of dough en-
closing a sweet filling. The formula uses a 40:60% blend of
hard and soft wheat flour plus sugar (60%), molasses solids
(10%), fresh yeast (5%), lard (30%), salt (1.25%) and water
(30%). The sweet filling is prepared by blending soft and hard
wheat flours (30:70%), sugar (30%), shortening or lard (35%),
fresh yeast (1%), whole eggs (30%), salt (1.25%) plus colorants
and cinnamon as desired. Dough ingredients are fully mixed,
the dough is divided into two equal parts, and one piece
shaped into a round thick layer. The sweet filling is spread on
this dough and the second layer of dough placed on top, seal-
ing the product. The surface of the top layer is decorated,
dusted with sugar, and perforated with a fork to prevent the
formation of air pockets during baking. After 1 hr of fermen-

tation the dough is baked at 200°C. (Anonymous 1990c).

Champurrada is a popular sweet, flat bread in Guatemala. It is round and flat; 10–15 cm in diameter and 0.5 cm thick. *Champurradas* are made by the straight-dough method, using hard wheat flour and the ingredients used for *pan dulce*. A significant difference is the short (10–15 min) fermentation time given to this dough (F. Saquimux, *personal comm.*).

Products Made from Soft Wheat

Cookies. Cookie consumption is associated mainly with urban food habits. However, these products are frequently pur-

Figure 2-6. *Bizcochos*, sweet breads of Mexico (Courtesy: El Mundo del Pan, Mexico).

chased by the rural population because they can be stored for extended periods of time. Mexican and Central American bakeries and/or cookie plants produce a variety of sugar cookies: plain, enriched with various ingredients (i.e. ground nuts, almonds, chocolate chips) and flavored. They are commonly prepared using soft wheat flour or the low protein flour streams from hard wheat milling. Formulas and production methods (e.g., wire cut, rotary mold), are typical of those in the USA (see Chapter 1).

Breads. In El Salvador, *pan dulce* is the name given to sweet breads prepared with a yeast-fermented stiff dough high in fat content, called *masa blanca*. A typical formula to prepare *pan dulce* dough is: soft wheat flour (100%), sugar (20%), shortening (40%), fresh yeast (2.5%), salt (1.25%) and water (15–25%). Although the dough is fermented, it does not increase in volume significantly, due to its low water content. The fermented dough is molded into a variety of shapes. Names given to the sweet breads made with *masa blanca* are: *cachitos, viejitas, punte,* and *zepelin* (Anonymous 1990c).

In Guatemala sweet breads, also known as *pan dulce*, are very popular in both rural and urban areas. The general formula differs slightly from that used in El Salvador. A typical formula consists of: soft wheat flour (100%), salt (1.25%), yeast (2.5%), sugar (27.5%), shortening (25%), baking powder (2.5%) and water (32.5%). Solids are dry blended, yeast and water are added and dough formed by hand-kneading for 8–12 min. Mixed dough is allowed to stand for 5 min before it is divided into small balls, rested for 10 to 20 min, shaped as desired, and proofed for 1.5 to 2 hr. Baking is at 275°C for 20 to 30 min (B. A. Cabrera, *personal comm.*). In general these breads have an aerated, semi-dense crumb. Popular names given to varieties of *pan dulce* include *concha, cubilete, bizcocho, moyete,* and *xeca*, (F. Saquimux, *personal comm.*).

Pan de pueblo is a sweet, whole-wheat bread made in the family bakeries of small towns of central Mexico. Its formula consists (percent whole meal weight basis) of: soft wheat whole meal flour (100%), ground molasses (40%), baking powder (0.6%), shortening (5%), fresh yeast (3%), whole eggs (2%) cinnamon (2%) and water (50–60%). Produced by the sponge and dough method, all the yeast, half of the flour and half of the water are mixed to a soft dough. This is allowed to ferment overnight to create the sponge. After fermentation, the rest of the flour, the molasses dissolved in warm water with cinnamon, and the rest of the ingredients, are added to the sponge and mixed to a smooth dough. The dough is fermented 30 min, divided and molded. The molded dough pieces are placed on greased trays, decorated with a soft shortening-sugar paste, allowed to proof for 60 min, and baked at 200°C for 20 min. The final product is a thick, dense bread (Fig. 2-7).

Tortilla de harina. The Guatemalan flour tortilla is a homemade product consumed mainly in those rural areas where wheat is grown. Its formula is different from those of the tortillas consumed in Mexico and the USA. The formula for Guatemalan flour tortilla consists of sifted soft wheat whole meal (454 g), whole eggs (20 g), sugar (120 g), shortening or oil (20 g), salt (1–2 g) and water (150 to 200 ml) sufficient to make a soft dough. All ingredients are mixed and kneaded to a soft, slightly extensible dough. The dough is divided into small balls, shaped into disks by hand or with a tortilla press, and baked on a hot plate. The tortilla is consumed plain to accompany meals (H. E. Carranza, R. Mora, V. Azañon, *personal comms.*).

Cakes - *Pastel de masa amarilla.* Salvadoran cakes called *pasteles* include pan-cakes (*pastel pacho*) and pie-like (*pastel de piña*) cakes. They are made with *masa amarilla*, which is *masa blanca*, the dough used for *pan dulce*, with yellow color added (Anonymous 1990c).

Quesadilla is a cheese pan-cake, popular in the rural areas of El Salvador. This product is made from a batter. A typical

Figure 2-7. Typical whole wheat sweet bread (*pan de pueblo*) of central Mexico.

formula consists of soft wheat flour (100%), sugar (50%), whole eggs (30%), fresh cheese and cream (variable). All of the ingredients except the eggs are blended, then the eggs, previously beaten, are incorporated into the mix to form a thick batter. The batter is placed in greased pans and baked at 350°C (Herrera 1982).

Products Made from Durum Wheat

A large variety of Italian-type pasta, made with durum wheat, is manufactured and consumed in Mexico. Formulations and production processes are similar to those used in the USA. From the various products thin-coil vermicelli, cooked in water and chicken or beef concentrate, is very popular in both urban and rural areas, where it is consumed at lunch as a soup.

Only small quantities of durum wheat is consumed in the CA region, exclusively in the form of Italian-type pasta products (both long and short goods). In CA countries, pasta products are urban foods consumed mainly in restaurants. Durum wheat is imported for the manufacture of pasta products only by Costa Rica and Guatemala (C. Salas-Fonseca, H. E. Carranza, R. Mora, V. Azañon, *personal comms.*). Belize imports all its durum pasta needs (Anonymous 1990a), while the rest of the countries import durum wheat pasta and/or use hard wheat for its manufacture (Barraclough 1984, M. López, *personal comm.*).

Other Products

Atole is a thick, sweet beverage popular in Guatemala. It is made by boiling ≈120 g of toasted wheat whole meal, ≈120 g of sugar and a stick of cinnamon in 2 L of water (or occasionally water and milk). Boiling is for 12–20 min, depending on the desired thickness (H. E. Carranza, R. Mora and V. Azañon, *personal comms.*).

Café de trigo. A hot wheat-coffee beverage, *café de trigo*, is popular in Guatemala. It includes a blend of 10% coffee and 90% wheat, toasted and brewed (H. E. Carranza, R. Mora and V. Azañon, *personal comms.*).

Acknowledgments

I wish to acknowledge the contributions in the form of personal communications and in other ways from the following persons in the preparation of this document: O. Espinoza-Chávez, Camara Nacional de la Industria Panificadora, Mexico.; C. A. Monroy, fabrica de pastas "La Moderna", Mexico; H. E. Carranza, R. Mora, V. Azañon and F. Saquimux, Instituto de Ciencia y Tecnología Agricola, Guatemala; M. López and L. Saad, CIMMYT, Mexico; R. Alfaro-Monge, Ministerio de Agricultura y Ganaderia de Costa Rica; F. Trejo-Ballesteros, Consejo Nacional de Producción, Costa Rica; C. Salas-Fonseca, Estación Experimental Fabio Baudrit M., Universidad de Costa Rica, Costa Rica; B. A. Cabrera, Guatemala; J. Abilio-Orellana, El Salvador; R. Arias-Díaz, Centro de Investigación y Estudios en Seguridad Social, Mexico.

References

Agurto-Vilchez, S. and Utting, P. 1985. La comercialización de alimentos básicos en Nicaragua. In Comercialización interna de los alimentos en América Latina, seminario internacional CIAT, Cali, Colombia, 11-13 Julio 1984. Scott, G. J. and Costello, M. G. eds. Ottawa, Ont., CIID, pp 253.

Anonymous. 1988. "Focus on Costa Rica". World Grain November/December, p 18.

Anonymous. 1989. Internal report. Departamento de Estudios Economicos, Departamento de proveeduria CNP y Planta Barranca, Costa Rica.

Anonymous. 1990a. "Belizean wheat flour mill supplies all the country's flour needs". World Grain, March, p. 28-29.

Anonymous. 1990b. Guatemala. World Grain, July/August, p. 26.

Anonymous. 1990c. El trigo; arraigo Latino-Americano. El Mundo del Pan (Mexico) 2(15):58-60, 62-65.

Anonymous. 1991a. El pan en America Latina: arte y costumbres ancestrales, " el alma de Guatemala". El Mundo del Pan (Mexico) 26(30):14-16, 18-19.

Anonymous. 1991b. Panes olvidados: Bizcocho. El Mundo del Pan 3(26): 30-32, 34-35.

Anonymous. 1991c. El Salvador. World Grain, May 1991, p. 29.

Barraclough, S. 1984. Un analisis preliminar del sistema alimentario de Nicaragua. Report no. 83.1, U. N.-UNRISD, Switzerland, pp 156.

Barranco-Fernandez, R. 1991. Producción y distribución del trigo en México. Revista Pan (Mexico) 38(450):11-12, 15-16, 18.

Food and Agriculture Organization of the United Nations, 1992, AGROSTAT, Rome.

F'uguemann, A. E. 1991. La industria molinera de trigo y su posición en el Tratado de Libre Comercio. El Mundo del Pan (Mexico) 3(24):49-51, 53.

Futrell, M. and Jones, R. 1982. Use of grain sorghum as food in Southern Honduras. In Proceedings of the grain sorghum quality workshop for Latin America. L. W. Rooney and V. Guiragossian, eds. CIMMYT, Mexico, pp 213.

Herrera, A. V. 1982. Usos del sorgo en alimentación humana en El Salvador. In Proceedings of the grain sorghum quality workshop for Latin America. L. W. Rooney and V. Guiragossian, eds. CIMMYT, Mexico, pp 213.

McWard, C. 1991. Mexico: country in crisis pulls itself out of debt and stagnaction with a hope of a North American Free Trade Agreement. World Grain, October, p. 6-7, 9-10.

Mielke, M. J. 1990. "El mercado del trigo mexicano y sus posibilidades comerciales". A translated summary of the original document written by the author for the Economic Research Service of the U. S. Department of Agriculture. Agro-Sintesis (Mexico), February, p. 30-31, 34-35, 38-39, 42-43, 46-47, 49-50.

Ministerio de Economía de Honduras. 1992. Internal report of the Department of Internal Commerce on flour production and consumption in Honduras.

Perts, G. A., James, A. W. and Pérez, J. 1981. La investigación sobre harinas compuestas en Nicaragua. Internal report. Laboratorio de Tecnología de Alimentos (LABAL), Ministerio de Agricultura, Managua, Nicaragua, pp 59.

Pomeranz Y. 1987. Modern Cereal Science and Technology, VCH Publishers Inc. New York, NY, pp 486.

Qarooni, J., Ponte, J. R. Jr., and Posner, E. S. 1993. Tortilla production with hard and soft white wheat flour. Assoc. Oper. millers. Tech. Bull. In press.

Serna-Saldivar, S. O., Rooney, L. W., and Waniska, R. D. 1988. Wheat flour tortilla production. Cereal Foods World 33: 855, 857-58, 860-62, 864.

U.S. Wheat Associates of Panama. 1992. El pan en Panamá. El Mundo del Pan (México) 4 (33):14-16, 18-24, 26-29.

CHAPTER 3

Wheat Usage in South America

José L. Robutti and Elsa de Sá Souza

Introduction

Wheat was introduced in South America by the European explorers (Chapter 2). The crop had to compete with others better adapted to the environment. It was only in the southern cone (the area comprising the countries of Argentina, Chile, Bolivia, Paraguay, Uruguay and southernmost Brazil), the most temperate part of the subcontinent, where wheat became important as a commercial crop. Argentina is the main producing country. Brazil and Chile also produce significant amounts of wheat (Table 3-1). However, Argentina is the only net exporter of the region. All the other countries have to import wheat in various amounts to meet their demands. Argentina is the only country that grows a small quantity of durum wheat (Table 3-1).

Per capita consumption varies widely depending on location, and ranges from 150 kg/year in Chile to 25 kg/year in Colombia. In this sense two different groups of countries exist. In the southern cone, either per capita (Chile, Argentina and Uruguay, whose populations have a strong European component) or total consumption (Brazil, in which wheat consumption is highly concentrated in the southern states where most of the European settlement has occurred) is high. If the Brazilian per capita consumption is assigned to this area, then it is going to be much higher than the commonly reported average. The second group consists of all the other countries. In this case, consumption ranges from intermediate (e.g., Bolivia) to

Table 3-1. Wheat Production in South America

	Country									
	Argentina	Chile	Brazil	Uruguay	Paraguay	Bolivia	Peru	Ecuador	Colombia	Venezuela
Estimated population 1987 ($\times 10^6$)	31.5	12.4	141.3	3.1	3.9	6.7	20.7	9.9	29.9	18.3
Production 1985/87 (1,000 tons/year)	9,167	1,555	5,353	262	264	79	115	18	75	...
Net imports 1985/87 (1,000 tons/year)	−6,000	220	3,338	60	38	305	989	243	661	1,063
Per capita utilization 1984/86 (kg/year)	111	150	56	132	72	68	54	27	25	59
% of area planted to spring bread wheat	99	48	100	100	100	N/A	N/A	N/A	N/A	N/A
% of area planted to winter bread wheat	0	25	0	0	0	N/A	N/A	N/A	N/A	N/A
% of area planted to facultative bread wheat	0	27	0	0	0	N/A	N/A	N/A	N/A	N/A
% of area planted to durum wheat	1	0	0	0	0	N/A	N/A	N/A	N/A	N/A

Data from: CIMMYT. 1989. CIMMYT 1987–88 Hechos y Tendencias Mundiales Relacionados con el Trigo. Nuevamente la Revolucion del Trigo: Tendencias Recientes y Retos Futuros. Mexico, D.F.:CIMMYT.

N/A = Not Available

very low (e.g., Ecuador). These countries did not have such a strong European settlement as those of the first group.

Wheat usage in South America by product category is roughly as follows: bread (70%), pasta (18%) and cookies (4%). The remainder goes to various miscellaneous uses.

Most of the wheat is processed to flour in large commercial mills. In the rural areas or small cities of many countries grain is still milled by communal stone mills similar to that shown in Figure 3-1.

Most of the bread consumed is of the French type and produced in small bakeries. The modernization level of small bakeries varies from country to country but, in general, is low. In many instances all or part of the bread making process is still performed by hand. This can be attributed to several reasons, among them: small bakeries of low production output, cheap labor costs, the high investment needed in machinery, the varied shapes that can be obtained by hand and resistance to change (linked to cultural issues). Modern baking plants enjoy lower production costs. This reduces price which, in turn, enhances consumer's demand for bread. Most South American bakeries turn out fresh bread from early morning till late evening thus maintaining a constant supply of attractive products. More recently supermarkets are becoming bread producers. This extends the market to a larger number of consumers. Most of the bread produced by supermarkets is of the French type. In many supermarkets consumers can observe the bread production process. Large bakeries which produce mostly pan bread are, in general, equipped with modern machinery.

The pasta industry, unlike the bread industry, has a generally high technological level. Pasta plants are large commercial units. The same is true for the cracker and cookie industry. Pastries, confectionery and sweet goods are produced primarily by small bakers, though a few large commercial plants exist in some countries.

Traditional and local products are of little significance from the quantitative point of view. Making cakes and cookies as well as pasta and empanadas at home is becoming less and less popular.

Wheat, Flour, and Flour Usage

ARGENTINA

Wheat was brought to the part of South America that is now Argentina by the Spanish explorers. However, it did not be-

Figure 3-1. Communal stone mill similar to the ones still in use in rural parts of South America. (Reprinted, by permission, from Escobari de Querejazu 1987)

come a commercial crop until late in the 19th century. By that time massive European immigration had started to settle in the pampas, resulting in a rapid increase in population of the sparsely inhabited region. The pampas comprise the east central plains of Argentina and have very favorable climate and soil conditions for grazing and crop production. Settlement modified the land use pattern from cattle ranching to a mixed pattern where farming and ranching shared the use of the land. By the 1920's wheat occupied 55% of Argentina's farm land. That figure is now about 35% (Coscia, 1984). In the 1920's Argentina contributed 5% of the world production and 20% of its exports. At present these figures are 2% and 5% respectively. However, the country is still the main producer in South America and the only net exporter (Table 3-1). As shown in Table 3-1, a small amount of durum wheat is grown as well. Recent personal communications (Torchelli, 1992, and Jensen, 1992) indicated that production of durum wheat has increased, to an estimated level of 50,000 to 100,000 tons/year.

Per capita wheat consumption in Argentina is about 125 kg/year (Coscia, 1984). Wheat is the number one source of calories and the number two source of protein, after beef. Products derived from wheat supply 31% of the 3,000-calorie average Argentine daily intake (Coscia, 1984).

Milling. The Argentine milling industry processes 3.9 million tons of wheat yearly into flour, farina and bran. Bran is used mainly as animal feed.

At present there are 95 flour mills having individual outputs ranging from 10 tons/day to 1,200 tons/day. In recent years most of the milling industry has modernized their equipment to the world standards. Modern mills have individual outputs ranging from 250 to 600 tons/day. An annual flour production of about 3 million tons is distributed among its various consumers, as shown in Table 3-2. The most important

Table 3-2. Flour Consumption by Use in Argentina

	Year			
	1987	**1988**	**1989**	**1990**
Uses	%	%	%	%
Bakeries	76	79	78	71
Pasta	6	7	7	8
Crackers and cookies	7	5	6	6
Retail sales	6	6	6	6
Exports	1	1	2	5
Others	4	2	1	2

Data from: Federacion Argentina de la Industria Molinera, 1992

use of flour is for bread making. This accounts for approximately 2 million tons/yr. Other uses such as pasta, cookies and crackers, and household consumption account for a much lower proportion. Though flour exports are a small percentage of total produced they increased significantly in 1990.

The Argentine Food Codex (AFC) classifies flours according to their ash content (Table 3-3). AFC also defines whole meal, bread wheat semola and bread wheat semolina. AFC does not classify flours according to their functional characteristics, nor does it classify according to their end uses. However, mills produce flours of differing functional characteristics, specifically for cookies and specialty breads. This is done by blending wheats or modifying flour rheological properties by means of physical or chemical treatments.

Millers as well as end users place great importance on wet gluten content (though test weight, dockage, and moisture and protein contents are also tested). It is determined by a washing technique using the Glutomatic equipment. When purchasing flour in bulk, its gluten content as well as its farinogram and/or extensogram and/or alveogram characteristics are taken as quality parameters.

No soft wheat cultivars are grown in Argentina. Therefore, for cookie and other traditional soft wheat industries, hard wheat flour must be modified either at the mill or while making the end product by mellowing the gluten with reducing agents or proteases.

Durum wheat is milled in the few mills where semolina is produced for pasta production.

Baking. According to the 1985 National Economic Census there are 13,345 bakeries (including those engaged exclusively in sweet goods production). This accounts for 1.1% of the whole Argentine manufacturing industrial output. Ninety five percent of the bakeries process daily up to 500 kg flour; 4.25%

Table 3-3. Maximum Moisture and Ash Contents in Flour Allowed by the Argentine Food Codex

Type of Flour	Moisture %	Ash % (d.b.)
0000	15.0	0.492
000	15.0	0.65
00	14.7	0.678
0	14.7	0.873
½0	14.5	1.35

Data from: Codigo Alimentario Argentino, 1992

between 500 and 2,500 and only 0.75% more than 2500 kg. Annual bread consumption is at about 70 kg per person. (Colomar, 1984).

Traditional (Small-Scale) Bakeries. Fresh bread is sold mainly at the bakery to a small neighborhood market. Recently there has been a tendency towards the selling of bread at other outlets such as groceries and supermarkets.

Most of this bread is of the French type, baked on the oven hearth with no pan. The commonly consumed bread comes in various sizes and shapes (*flauta, flautita, mignon, felipe*, etc) shown in Figure 3-2. *Mignons* weigh on the average 35 g and *flautas* 100 g per loaf. All of them are produced from "000" flour. The process most often employed is the straight dough (Chapter 1) in which all ingredients are mixed together in one step. Mixing is generally long because of low mixer efficiency. The dough resulting from this type of mixer is not yet optimally developed. Therefore, it is passed through a sheeter for further development. "Developed" dough is left to rest in bulk on the working table (the *torno*) for 15 minutes. Afterwards, the sheet is divided into sticks and these are cut into pieces of the desired final weight. Pieces are formed by sheeting and rolling and left to rise till the initial volume is doubled. Just before baking cuts are made on the proofed dough's top surface. Either straight or angled, they are characteristic of French-type bread. A typical formula contains flour, salt, yeast, water and potassium bromate (up to 70 ppm). It may contain some malt flour or extract. The proportion of yeast is about 2% and that of salt 1.5% of the flour.

A variant of this process, also used in Argentina, is the *tablas* method. It is a straight dough process in which the amount of yeast in the formula is reduced to approximately 0.5%. After mixing and sheeting, the dough is divided, covered with a piece of cloth and the pieces are left to rise on wooden boards (*tablas*) for several hours. The longer fermentation time counteracts the reduced yeast level. Because it is so lengthy (12 hours), the process takes place during the night with loaves baked early the next morning. A flow chart of the process can be seen in Figure 3-3.

As mentioned above, traditional bakeries operate at a low technological level; specifically, this means slow mixers, sheeters, molders, fermentation rooms and brick ovens heated by natural gas, diesel fuel or wood. Additionally, ovens are so large that the surface area required to build a traditional bakery is considerable. There is a trend towards the replacement

Figure 3-2. Different types of bread consumed in Argentina.

and upgrading of traditional bakery equipment with rapid mixers, sheeters, cutter-molders, fermentation cabinets and rack ovens, which occupy a much smaller area. An example of this can be seen in Figure 3-4. This allows production schedules to offer fresh bread and other specialties throughout the day.

In recent years, the consumption of baked products with a higher fiber content has increased. The same is true of the use of composite flours in baked products. Bran bread is produced with inclusion of 10–15% wheat bran. Likewise whole wheat, rye, oat and/or oat bran containing breads are now produced more commonly. Also consumed are small amounts of bread made with flour to which defatted and toasted soybean meal has been added. In all these cases loaves are generally 300–400 g in weight but small rolls can also be found. Specialties also include breads made with milk that show a very fine crumb structure (such as *pebetes*), hot dog rolls, hamburger buns, *figasas* and *figasitas*, and rolls flavored with cheese, onion and other flavors.

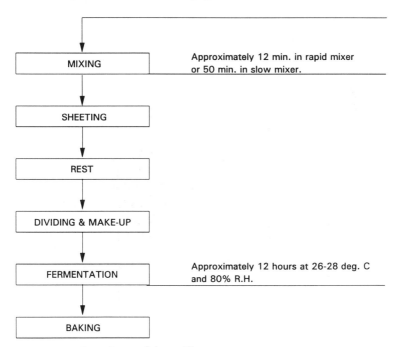

Flour, water, yeast (0,5%), salt, potassium bromate, eventually malt flour and other dough conditioners and reducing agents.

MIXING — Approximately 12 min. in rapid mixer or 50 min. in slow mixer.

SHEETING

REST

DIVIDING & MAKE-UP

FERMENTATION — Approximately 12 hours at 26-28 deg. C and 80% R.H.

BAKING

Figure 3-3. Flow chart of the *tablas* process.

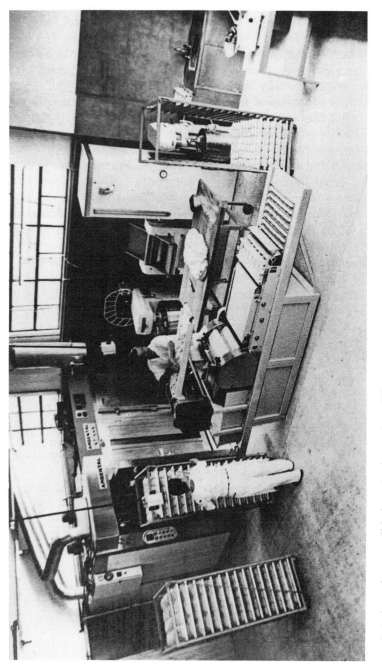

Figure 3-4. A modern small bakery in Argentina. (Photo courtesy Argental S.A. de Argentina. Used by permission.)

Galletas are made from semola (fine semolina). Among formula variations for *galletas* are those with added bran, no salt, etc. *Grisines* are crisp bread sticks ranging in size from 10 to 20 cm long and 1 to 2 cm in diameter.

Bakeries produce confectionery products either by biological or chemical leavening. Among the former are *facturas* (Fig. 3-5), produced daily and sold by the dozen, and *pan dulce* or *panettone*. *Panettone* is a seasonal product customarily eaten at Christmas and New Year's Day dinners. Among the latter category are cakes, pies and the so-called *masas secas* and *masas con crema*. These are small pieces of different types of batter with different fillings and icings. *Facturas* are produced using yeast-leavened dough and contain a high level of fat. This results in a flaky structure. *Facturas* may or may not be filled with different types of jam or with *crema pastelera*, made from milk, flour, eggs and sugar. The most common *factura* is the *media luna*.

Also sold at bakeries are prebaked pizza-disks. These are made from a yeast-leavened dough, which is sheeted in a sheeting machine, divided in pieces, hand-rounded and molded, rested and baked. Before baking the surface is covered with tomato sauce.

Large-Scale Industrial Bakeries. The major product of large-scale bakeries is pan bread in either whole or sliced loaves. Such large bakeries are located in and around Buenos Aires, as well as in some other large cities of the interior. The proportion of pan bread in the bread market increases as more and more women join the workforce and commuting time increases. Both facts make the daily purchase of fresh French bread very difficult. Pan bread also is suitable for sandwiches and other quick and easy-to-fix meals.

Both the straight and the sponge and dough mixing processes are used for making pan bread at the large industrial level. A typical formulation is as follows:

Ingredient	% (Flour weight basis)
Flour	100
Water	65 (variable)
Yeast	2
Yeast food	0.25–0.50
Salt	2
Sugar	6
Skim milk solids	6
Shortening	4

Figure 3-5. Typical pastries consumed in Argentina.

A number of minor ingredients are often included in the formula, primarily to alter the dough rheological properties. These include: fungal amylases, less frequently fungal proteases, cysteine or sodium metabisulfite, ascorbic acid, potassium bromate, sodium stearoyl-lactilate (SSL), mono and diglycerides (45%) or concentrated α–monoglycerides.

Typically, Argentine flours have long mixing times and produce strong elastic doughs. These rheological characteristics make it necessary to use reducing agents in order to decrease mixing time and obtain a homogeneous dough capable of being adequately sheeted and molded. Without them an elastic and bucky dough results.

Article 747 of the AFC allows the addition of calcium and/or sodium propionate as an anti-mold agent in products with 12% moisture or more without stating it on the label. Amounts allowed by product are as follows:

White flour products: up to 0.32% flour weight

Whole meals, mixtures thereof and mixtures with white flour: up to 0.38% of whole meal or mixture.

Article 750 lists all of the permitted emulsifiers while article 751 allows (no label statement required) one of the following:

a) Sodium or potassium bromate up to 7 g/100 kg flour

b) Azodicarbamide up to 4.3 g/100 kg flour

c) l-Ascorbic acid up to 200 mg/kg flour.

In the straight dough process all ingredients are mixed together and left to ferment. The first punch takes place once 60% of the fermentation time has elapsed. The punching procedure is repeated till a soft dough lending itself to sheeting and molding is obtained.

In the sponge and dough process a portion of the flour is mixed with the yeast, yeast food, shortening, water and sugar and allowed to ferment 4–6 hours at 26°C and 70% R.H. When fermentation is complete the rest of the ingredients are added and mixed to a dough. The dough is then allowed to relax for 15–30 min before proceeding to make-up stage (division, rounding, intermediate proofing and molding).

After panning and for both processes, dough pieces proof for 50–70 min at 35–39°C and 80–85% R.H. Baking is generally done in traveling band ovens at 215–230°C for 20–30 min or longer depending on size, shape and type of product.

A broad array of pan breads are produced. The list includes milk bread, *pan salado de mesa* (eaten with meals and having a higher salt content), whole meal, bran, soybean meal, rye, and corn breads. Packaged products such as hamburger buns, hot dog rolls, bran rolls, prebaked pizza crusts, *panettone* and *alfajores* are also produced by large-scale industrial bakeries.

A unique product made by both small and large-scale industrial bakeries is sandwich or English or crumb bread. Bread for the "crumb sandwiches," popular in the cities, is made in large loaves (3–12 kg) from high gluten "0000" flour (or mixtures of "0000" and "000" flours), yeast, shortening, milk and/or water, emulsifiers, dough conditioners, oxidizing agents and preservatives. Because the baking pan is rectangular, with a lid, the loaf produces square slices with a thin upper crust. Slices are cut 3–5 mm thick, and the crust is completely removed. These slices, in turn, are cut into rectangles of approximately 10 cm long and 5 cm wide. Butter or mayonnaise is spread on either slice and the sandwich filled with a variety of products (ham, cheese, tomato, lettuce, peaches, pineapple, tuna, etc).

Crackers and cookies. The cracker and cookie industry began in Argentina at the end of the 19th century. Now it consists of large factories with modern equipment, most of it of European origin. There are also small and medium size establishments equipped with production lines that are either Argentine or European-made.

Both crackers and cookies are produced; the former being consumed in larger amounts. Recently crackers have become competitors to bread in the marketplace.

The production of saltine crackers starts with a sponge made from flour, yeast and salt. The sponge ferments in bulk at 27° C and 80% R.H for 12 to 18 hours, then the remaining flour, fat and enzymes are added. Fermentation continues for a time of 18–24 hours. A typical formula is as follows (% flour basis):

Sponge

Flour	33.0
Fresh yeast	0.5–1.0
Water	13.3

Addition at dough up

Flour	67.0
Shortening	19.0
Sodium bicarbonate	0.3
Salt	1.0
Malt extract (non diastatic)	0.7
Water (approximately)	20.0

Sheeting, lamination, docking and baking are similar to North American saltine production discussed in Chapter 1.

A variant to saltine crackers, specially produced as snacks, are flavored (savory) crackers. They are sprinkled with salt and sprayed with fat or oil after baking. The dough is either yeast-leavened and subjected to sheeting or chemically developed, as in the case of semisweet doughs. In general these products have a tender and porous texture due to the use of ammonium bicarbonate.

Cookies come in a broad variety of shapes, sizes and flavors. The dough is generally made by a two-stage mixing procedure which first creams the fat and sugar before the addition of flour and other ingredients and mixing to a dough. A typical formula is:

Ingredient	% (Flour weight basis)
Flour	100
Sugar, syrup and/or glucose	30
Fat	21–25
Water	4–5
Leavening Agents	variable
Flavorings Material	variable
Colorants	variable

Most cookies are produced by rotary molding, extrusion and wire cutting or by extrusion and reciprocating cutting. Very few types are produced by sheeting, gauging and cutting.

Puff pastries are very popular in Argentina. Shaped as sticks or as palmiers, they are made from doughs subjected to folding and sheeting with layers of shortening between layers of dough. During baking the layers separate giving the desired flaky structure. The surface is sprinkled with sugar before baking to give a glossy appearance. Fats used—of a high melting point—are either vegetable shortening or mixtures thereof and refined beef fat.

Pasta products. Several types of products fall under this broad designation. They can, in fact, be produced from different raw materials (flour, durum semolina, and break middlings) and have various sizes and shapes. Two types of pasta exist in the market: dry (with a moisture content of less than 12%) and fresh pasta (with up to 35% moisture).

Fresh pasta is produced in factories where the entire process can be watched by customers from the sales counter. There are 698 of these plants, most of them (443) located in and around Buenos Aires. The product is sold at the point of production and supplies the demand of a small neighborhood market. This type of pasta must be sold within 24 hours after being produced as required by article 720 of the AFC.

In recent years, long-life fresh pasta has been marketed in Argentina. This type of pasta, manufactured in large and medium size factories, is pasteurized and packaged in a modified atmosphere to maintain shelf life and includes products with fillings such as ravioli and capelletti. Frozen fresh pasta has recently appeared in the market.

The annual Argentine output of 230,000 tons of dry pasta is produced by 169 plants located throughout the country. Per capita consumption is estimated at 7.5 kg/year. Large plants, those with an output capacity between 150 and 200 tons/day, are equipped primarily with Italian equipment imported in 1970's. Some of that equipment was upgraded or replaced during the last five years. Medium-sized plants produce 30–40 tons/day and small plants produce 5 to 10 tons/day. In general large and medium-sized operations employ the continuous production processes very similar to those used in other parts of the world. (see Chapter 1).

Manufacturers classify their production as long goods and short cuts, according to size and shape. The AFC defines several types of pasta products according to the process (extrusion or sheeting) by which they were obtained. The AFC also defines pasta products (*fideos*) according to the raw materials used. *Fideos de semola* are those made from 100% durum semolina, *fideos semolados* are made from a 50% mixture of durum semolina and bread wheat flour or break middlings. The term *fideos* with no qualification designates those made from bread wheat flour.

According to the AFC regulations, dry pasta must have a moisture content <14% (usually it is <12%), and acidity (as lactic acid) must be no higher than 0.45%. Mixtures of mono and diglycerides and/or with high concentration monoglycerides can be added as well as vegetable colorants such as ß-carotene, turmeric, saffron and others. Fresh pasta may contain sodium and/or calcium propionate (max. 0.25%) or sorbic acid and/or sodium, potassium or calcium sorbates (max. 0.05%).

Dry egg macaroni must contain no less than two yolks per kg of semolina, flour or their mixtures. Cholesterol content can not be less than 0.04% (dry basis). Fresh egg pasta must contain 3 yolks per kg of dough, and its cholesterol content has to be equal to or higher than 0.06% (dry basis).

Traditional products. Empanadas (Fig. 3-6). The dough for empanadas is made of wheat flour, beef fat, salt and water. Round, flat dough pieces (about 10 cm diameter and 2 mm thick) are formed and a filling placed at the center. The disk is folded along its diameter and the edges crimped one against the other to create the final shape. Empanadas are then either baked or deep fried in beef fat. Fillings are made generally with ground beef, onions, peppers, olives, prunes, raisins, etc. and a variety of seasonings. The filling ingredients and the cooking process (frying or baking) mark the differences among empanadas from different regions. Empanadas were originally homemade products. However, packaged empanada shells have been sold commercially for nearly 40 years. A dough is mixed from "0000" flour (100 parts by weight), water (40), salt (2.2) and preservatives (0.3) and then sheeted. A layer of empaste (20 parts flour and 50 parts beef fat either alone or mixed with shortening) is sandwiched between two layers of dough, sheeted, folded and sheeted several times more. Disks are cut from the flat dough and packaged in 12-unit packages. Disks can be cut of a larger diameter and sold in pairs as pie shells. These products required refrigerated storage.

To produce *pastelitos* a portion of quince jam is placed be-

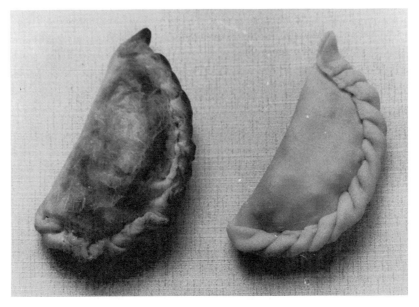

Figure 3-6. Empanada dough and fried product.

tween two square empanada crusts, the unit is given a flower like shape, fried in either beef fat or vegetable oil and sprinkled with sugar.

Tortas fritas are a homemade flour-based product. Pieces of a dough similar in formulation to that for empanadas are made to approximately 10 cm in diameter and 0.5 cm thick and deep fried in beef fat. Once fried, *tortas fritas* are sprinkled with either salt or sugar.

Alfajores are two flat cookies 5 cm in diameter and 0.5 cm thick sandwiching fillings such as *dulce de leche* (milk and sugar evaporated to a pale thick fudge), fruit jams, chocolate mousse, etc. and covered with different icings (chocolate, sugar, etc.). *Alfajores* originated as homemade products, though it is known that they were commercially produced in certain parts of Argentina as early as 1850. Normally, cookie manufacturers turn out a great variety of *alfajores* by modifying cookies, fillings and icings.

Galleta de campo. These products are produced at home in rural areas from a basic dough of flour, salt, baking powder or yeast, water and beef fat. The resulting dough is mixed, sheeted, folded several times and cut to a round or square shape. The baked piece is about 4 cm high, has a somewhat friable crust and a compact and flavorful crumb. The high fat content helps to extend freshness of the product.

Miscellaneous uses. Miscellaneous uses of wheat include self-rising flour for household consumption, cake and pie pre-mixes, modified flour as water binder especially in processed-meat manufacturing, starch and gluten production, wheat flakes, snacks such as pretzels, *palitos salados* (salted sticks) and others.

CHILE

A long-time wheat producer, Chile has the highest per capita consumption of wheat foods in the hemisphere. Per capita flour consumption totaled 83 kg in 1985 compared to the record highest of 109 kg in 1973. This consumption level is substantially higher than that of most other South American countries. In general, Chile supplies its needs with its own domestic production. However, during 1978–82 domestic wheat production dropped dramatically, forcing wheat imports, primarily from the United States. The economic troubles in 1981–82 prompted a sharp reversal in the governmental agricultural policies so that today, virtually all of the wheat milled in Chile is domestic (Sosland, 1988a).

The 123 large mills in Chile, on the average, produced a little more than one million tons of flour in 1985. Individual mills range in size from 300 tons/day to only 3.5 tons/day. At present the mills produce two grades of flour: *harina flor* (fine flour), a bakery type of 78% extraction and small amounts of flour with an extraction rate of 74–76% for special types of bread. Standards are set by a government agency which specifies a maximum of 15% moisture, 0.645% ash and a minimum of 9% protein. The official extraction rate set for flour is at 78%, the rheological standard (*cifra valorimetrica*) at 40, maltose at 1.10 and the sedimentation level at 25.

Chileans have long been fond of good bread. In the 1970's the government made efforts to upgrade bread quality. French bakers were brought in to introduce baguettes and other sophisticated baked products to the marketplace, with the result that a taste for high-quality bread was developed. As a consequence the milling industry has been compelled to supply their customers with flours that meet high standards. As in many European countries the principal flour buyers are small bakers. In fact there are only three large commercial bakeries in Chile. Although bakers attach some importance to flour color, more attention is paid to strength and water absorption (parameters more directly affecting earnings). The principal bread product in the country is a French-type bread called *mariquita*. Bakers also produce *hallulla*, a round flat product, resembling a tortilla but somewhat thicker (Sosland, 1988a). *Hallulla* is made from ground toasted whole wheat, salt and water. After mixing and kneading, round and flat pieces are formed and cooked quickly on hot stones or in a hot oven. The product puffs during baking to form a hollow interior. The layers can be cut open and the hollow filled with vegetables, meat and legumes (Vega, 1992, *personal communication*). There are governmental enrichment specifications. Potassium bromate is restricted to no more than 5 g per 100 kg. Despite the low protein content of flour few bakers use vital gluten because of the cost (Sosland, 1988a).

A sizable share of the flour milled in Chile is sold to cracker baking operations, which unlike most of the bread bakers are large industrial units (Sosland, 1988a).

Durum flour is utilized by pasta manufacturers. Some of the manufacturers are vertically integrated. There is one complex that turns out not only flour but pasta, crackers and commercial feed as well (Sosland, 1988a).

Bran (millfeed or offal) is sold both to domestic and foreign users. Some mills pellet their bran and export it to Europe (Sosland, 1988a).

Mote is much like that in Bolivia. It is prepared at home but also available in stores. Empanadas, as well as *alfajores*, are popular products. Their formulae and production differ little from their Argentine counterparts (Vega, 1992, *personal communication*).

URUGUAY

Uruguay is one of the highest per capita consumers of wheat in South America, though due to its small population, the total amount needed to meet the country's requirements is not large.

During 1988/91 an annual average of 360,000 tons of wheat were milled by the eight mills in the country. Flour produced in 1988 was 255,000 tons with an extraction rate of 73%.

In 1988, about 253,000 tons of flour were sold in the Uruguayan domestic market. The users of the flour were as follows: bakeries 69%, retail sales 15%, dry pasta 14% and fresh pasta 2%. Clearly most of the flour was used for baking purposes. As in other South American countries most of the bread is of the French type and produced by small bakeries. Pasta and cookies, on the other hand, are manufactured by large plants.

Empanadas and *alfajores* are popular in Uruguay. Their production and characteristics are much the same as those produced in Argentina (see above).

BRAZIL

With an annual consumption of 8 million tons, Brazil is the country with the highest usage of wheat in South America. Because of its large population, however, per capita consumption is relatively low. Domestic production supplies, on the average, one half of total consumption. Imports come mainly from the United States and Argentina. The total wheat milling capacity of Brazil is unofficially estimated between 9.5 and 10 million tons/year, making it the largest South American milling industry. The number of mills in operation is estimated to be between 140 and 180 units. Flour extraction rate varies geographically from 78% in the north, where primarily North American wheats are milled, to 70% in the south, where millers utilize primarily native and Argentine wheats. In most of the country two grades of flour are available: a special grade with a maximum of 0.5% ash and a common grade averaging

0.75% ash. Flour color is a major competitive factor, as well as gluten strength.

Brazilian government food regulations are published by the Associaçâo Brasileira das Industrias da Alimentaçâo when first established, or whenever changed or updated. These regulations define whole, common, and special flours according to extraction rate and ash content. Semola and semolina are defined according to particle size. Limits are set for microbial counts as well as for gluten content, acidity, moisture and ash content.

On the average bakers use 45% of the flour milled in Brazil. Roughly 30% goes to pasta and cookie manufacturers and 15% is for household use.

Most of the small bakeries produce a bread resembling French baguette varieties. There are approximately 25,000 bakeries in Brazil of which only three are large commercial units. Because of a sharp expansion in consumption during a recent period of wheat subsidy, bakers have been hard pressed to meet the increased demand for the subsidized wheat products. As a consequence many bakers adopted shorter time baking production procedures to enable them to turn out fresh bread on demand. Thus, the Chorleywood process has become popular in parts of the country. Ascorbic acid is used occasionally in both normal and continuous mix process but the use of potassium bromate is prohibited by law (Sosland, 1988b).

The six large pasta manufacturing plants use the same grade of flour as bread bakers. It is said that this industry has been able to develop a special technology to overcome the small granulation of bread flour, allowing it to produce high-quality pasta from conventional bread flour (Sosland, 1988b).

Brazilian food regulations also define different types of bread, cookies and pasta products and set specific limits for microbial counts, acidity, and protein, starch, moisture and ash (NaCl excluded) contents. For common bread, maximum moisture and minimum protein are 30% and 11.4% respectively.

PARAGUAY

Paraguay mills a yearly average of 270,000 tons of wheat in about 20 mills, most of them located in and around Asunción, the largest and capital city (Cáceres, 1992, *personal communication*). Most of wheat flour is used for production of bread. The most common type is the *galleta paraguaya,* which comes in various sizes: from a loaf with the diameter of a large orange to that of a small walnut (the latter being called *coquito*).

Galleta has a very low moisture content, a relatively high fat content and an extended shelf life of 15 days or even longer. Because most of the population lives in the country, large-scale production and consumption of French-type bread is rare. However, many small as well as a few large bakeries produce French-type bread in Asunción. Very low labor costs do not encourage investments in mechanization of these operations (Méndez, 1992, *personal communication*).

BOLIVIA

Per capita consumption of wheat in Bolivia increased significantly from 31 kg/year in 1950 to 88 in 1988 (Prudencio Böhrt, 1991). This is in spite of the fact that domestic production supplies only 12% of total needs. The proportions of domestic wheat devoted to different uses are illustrated in Table 3-4. Milling industry uses a relatively low proportion of the wheat, whereas most is used for household consumption as *mate, tostado* and other traditional products. A relatively high proportion of this use goes to *chicha* production. *Mate* is made by soaking intact grain with ash or lime followed by grinding and partial dehulling. The grinding operation is performed on the *batán*, a kind of saddlestone and *metate*. The resulting *mate* is finely ground and brought to boil with water. The mash is then drained through a cloth and the liquid phase (called *chicha*) is consumed as a drink either per se or with added sugar. The liquid phase can also be fermented by adding yeast. After a short fermentation an alcoholic beverage is made that is also called *chicha*. Although a certain amount of *chicha* is made from wheat in Bolivia, most of the *chicha* consumed in the Andean countries of South America is commonly made from maize. *Tostado* is wheat grain roasted on a hot plate or in an oven till it reaches a golden brown color. Tostado is eaten in a manner similar to roasted peanuts (Isnado, Gu-

Table 3-4. Usage of the Domestic Wheat Production (%) in Bolivia (1970–1986)

Utilization	1970	1977	1983	1986
Milling industry	30.0	26.0	21.0	19.3
Small mills	31.0	...	10.0	12.8
Chicha	7.0	...	17.0	...
Household consumption[a]	19.0	64.0[b]	40.0	42.5
Seed	13.0	10.0	12.0	10.0
Other[c]	...	...	...	15.4

[a] Consumed as trigo mote, tostado, etc.
[b] Chicha production and utilization in rural mills.
[c] Storage and other nonspecified losses.

Source: Prudencio Böhrt, 1991. Used with permission.

tiérrez, and Scott, 1992, *personal communications*). If tostado is ground and mixed with powdered sugar, the mixture is known as *pito*. If whole tostado kernels are cooked with water, salt and butter the resulting concoction is called *phyry* (Escobari de Querejazu, 1987).

Commercial wheat milling started in Bolivia in 1929. The industry operated with traditional methods until 1965 when simpler and automated methods for flour production were incorporated. Currently there are 18 mills with capacities ranging between 10 and 330 tons/day. Bolivian mills rely heavily on imports (94%) which come mainly from the United States and Argentina. Flour extraction rate is 76% and flour consumption between 200,000 to 250,000 tons/year (Escobari de Querejazu, 1987). The main uses of flour are shown in Table 3-5.

Bread in Bolivia is made mainly (90%) in small bakeries producing *pan de batalla* (battle bread). In addition, there are local types of bread such as *k'hasi*, *kaukas* (with added fat), *chakoso* and others. Industrial bakeries are few, producing a higher-quality bread. The output of small bakeries ranges from 200 to 500 kg/day while that of industrial ones is between 1,500 and 2,000 kg/day. In small bakeries, most of the bread-making operations are performed by hand. Fermentation is mostly inadequate. Baking is carried out in ovens heated by natural gas, diesel fuel or wood (Prudencio Böhrt, 1991). Cookie and cracker manufacture is carried out by a few modern plants. These plants have modern equipment for mixing, molding, cutting, filling and baking. Main ingredients are: flour, water, milk powder, eggs, cocoa, etc. Other ingredients are: butter, yeast, egg yolks and starch (Prudencio Böhrt, 1991).

Two groups of pasta manufacturers exist in Bolivia. The first one consists of plants equipped with old machinery and low production capacity. The second one consists of modern plants with large production capacities. Pasta is made from the same type of flour as used for bread and cookies. Usual ingredients are: flour, water and, occasionally, coloring materials. In some cases semolina is used. Enriched pasta is not produced in Bolivia, but eggs, powdered spinach, and other

Table 3-5. Evolution of Wheat Flour Utilization (%) in Bolivia (1978–1989)

Utilization	1978	1984	1989
Bread baking	78.5	75.0	78.0
Pasta	18.0	21.0	18.0
Crackers and cookies	3.5	4.0	4.0

Source: Prudencio Böhrt, 1991. Used with permission.

flavors are added as specialty items (Prudencio Böhrt, 1991).

Miscellaneous uses of flour include empanadas, and *salteñas* production. Empanadas are similar to the Argentine types. They may be colored with different edible vegetable colorants. Fillings are made of ground meat, onions, cheese, peppers, etc. and mixtures thereof. *Salteñas* differ from empanadas in the cooking process. The former are baked, while the latter are fried. *Sopapillas* are very similar to Argentine *tortas fritas* (Isnado, Gutiérrez and Scott, 1992, *personal communications*).

PERU

Peru imports an average of 900,000 tons/year of wheat mostly from Argentina, with a small tonnage coming from the United States. In 1990, the figure went down to 600,000 tons. Domestic Peruvian wheat, which has been grown since colonial times, has never played a significant role in supplying the milling industry and accounts for less than 5% of the grist (Anonymous, 1991).

According to millers' reports there are 18 mills in Perú with a total annual capacity of nearly 2 million tons. Originally, three grades of flour were produced: popular at 87% extraction, extra at 82% extraction and special at 75% extraction. Popular grade virtually disappeared after 1981 because of lack of quality and consumer acceptance (Anonymous, 1991).

Mills in Perú sell their flour directly to bakeries. Perú's baking industry consists of thousands of small and medium size bakers producing bread mostly of the French type in sizes ranging from 40 to 80 g loaves and consumed mainly in the cities (Méndez, 1992, *personal communication*).

Cookies are produced in Perú. Some of the larger cookies producers are owned by millers.

Mate and tostado are also produced in Peru. Mate is also known as *f'ata* and tostado as *cancha. Trigo pelado* results from cooking whole kernels in an alkaline solution (wood ashes). After cooking the kernels are cooled and bran is partially removed. The kernels are coarsely ground and consumed as similarly to rice (Jara and Chávez, 1992, *personal communications*).

ECUADOR

Flour consumption has soared in Ecuador, especially during the 1972–80 period, when it showed an annual increase averaging 7%. At the same time wheat production decreased sharply. Consequently, imported wheat accounts for 96% of the grist. At present Ecuador imports all of its wheat, usually No. 2 hard red winter with 12% protein, from the United States. Some durum and soft wheat are also imported as requested by the local pasta and cookie industries. In 1990, wheat was used primarily for bakery products (63%), with pasta products accounting for 15%, cookies for 16%, shrimp feed for 8% and other uses for 8% (Sosland, 1988c).

A total of 24 mills in the country have individual milling capacities ranging from 10,000 to 120,000 tons/year. Several types of flour are produced including enriched (addition of vitamins), bleached and improved. Flour extraction rate is generally 74–75%. Low-grade flour is used as filler in other food products, to feed shrimp, and other uses. It usually has as much as 2% ash. Ecuador's flour production totaled 290,000 tons in 1986. Per capita consumption averages 30 kg/year, well below the average of most other South American countries. This is due, in large part, to the fact that Ecuador's large Indian population consumes primarily maize and maize-derived products (Sosland, 1988c).

Bread consumption has become increasingly popular in the large urban centers where it has proved an effective method of feeding the population. The fact that many women in the cities take jobs in contrast to their traditional role in the countryside has also stimulated bread consumption. According to millers' estimates there are approximately 4,000 bakeries in Quito alone. Ecuadorian bakers produce a French-style baguette and have become increasingly demanding with respect to flour quality. Most of these plants utilize between 100 and 200 kg of flour daily. Total annual bread production has been estimated at 180,000 tons (Sosland, 1988c). A substantial proportion of the bread produced has sugar and/or fat and/or colorants, etc. added which makes it more expensive than just plain bread. This acts as a hindrance in expanding the demand for bread (Méndez, 1992, *personal communication*)

Although cookie and cracker production is only about one fifth that of bread, several large plants are in operation. The largest plant producing cookies and crackers utilizes 100 tons flour per month.

A sizable share of the flour marketed in Ecuador is for family use. Home-baked cookies and cakes are quite popular. A type of cookie called *hallulla* with a high lard content is quite commonly consumed in certain mountain areas. This *hallulla* is completely different from the Chilean one (Méndez, 1992, *personal communication*).

As a leading cultured shrimp exporter, Ecuador is expanding the use of low-grade flour in this application (Sosland, 1988c). Added to pellet shrimp feed it serves as a binder.

COLOMBIA

Wheat is not a major crop in Colombia (Table 3-1). The estimated production for 1989 is about 80,000 tons. Colombian wheat is classified as soft and has to be mixed with imported wheat for milling to bread flour.

Chuchuco, a product similar to mate, is prepared at home and available in stores as well (Vallejo, 1992, *personal communication*).

The highest demand for French-type bread is in Bogotá and some other large cities.

A sort of flat pancake similar to corn tortillas is made from flour in the wheat growing areas of Colombia (Vallejo, 1992, *personal communication*). About 10,000 tons of wheat is used as feed annually.

VENEZUELA

Less than 3,000 tons of wheat are grown in Venezuela annually, but the crop is an important food grain with about 1.2 million tons imported annually.

Mills produce 850,000 tons of flour a year for bakers, pasta manufacturers, the cookie and confectionery industry and for household use. Wheat-based products include French-type bread, a broad array of croissants, pastries, and pasta products (Anonymous, 1989). French-type bread consumption is rapidly expanding in Venezuela. Most bakeries are small establishments, owned and/or managed by Spanish or Portuguese immigrants. Quality is generally high and loaf size small (Méndez, 1992, *personal communication*).

Wheat products are increasingly substituted for rice and corn. The government of Venezuela has attempted to reverse that trend by requiring millers to incorporate 10% rice and corn flour in all wheat products. This has caused technical problems for both millers and bakers. Although the requirement is still in force, the availability of the nondiluted wheat flour has been scarce for the past few years (Anonymous, 1989).

A number of changes are expected in wheat usage such as: higher rates of blending different wheat classes, higher use of winter and/or softer wheats, lower bread but higher pasta consumption (Anonymous, 1989).

Acknowledgments

The authors wish to acknowledge their gratitude to the following persons:

Menella, D. and Fabrizio, C. from Bagley S.A.

Coppari, L.A. from Matarazzo SAIC.

Collingwood, W. from Federación de Industriales Fideeros de la República Argentina.

López Naguil, A. from Morixe Hnos. SACI.

Guarienti, E. and Rodriguez, O. from Embrapa for furnishing information on Brazil.

Troche, L. from INIA and Villamarzo, F. from Dirección Nacional de Granos for information on Uruguay.

Britos, U. from IAP for information on Paraguay.

Méndez, M. from Argental for general information on South America.

van Becelaere, N. for the drawings and Righi, H. for some of the photographs illustrating this chapter.

Colomar, L. and Añón, M.C. for reviewing this manuscript and giving helpful suggestions.

Palacios, Jorge from Buhler S.A.

Maiola, E. who reviewed the style and grammar of the English in which this chapter was written.

References

Anonymous. 1984. Plan de desarrollo sectorial 1984–1987: Agroindustria molinera. Ministerio de Industria, Comercio y Turismo. La Paz. Bolivia.

Anonymous. 1989. Focus on Venezuela. World Grain. 7(4):23.

Anonymous. 1991. Peruvian flour market liberalized. World Grain. 9(6): 42.

Colomar, L. 1984. Máquinas para la fabricación de pan en la Argentina. In: Congreso Mundial de Tecnología de Alimentos. Buenos Aires. 20–24 de setiembre.

Coscia, A. 1984. La economía del trigo. Hemisferio Sur. Buenos Aires.

Escobari de Querejazu, L. 1987. Historia de la industria molinera boliviana. ADIM. La Paz.

Prudencio Böhrt, J. 1991. La inseguridad alimentaria en Bolivia. El caso del trigo. ILDIS. La Paz.

Sosland, N. 1988a. Chilean millers adjust for free market. World Grain. 6(2): 8.

Sosland, N. 1988b. Brazil adjusts to no wheat subsidy. World Grain. 6(7): 16.

Sosland, N. 1988c. Flour milling in Ecuador. World Grain. 6(11): 9.

CHAPTER 4

Wheat Usage in Western Europe

Wilfried Seibel

Introduction

Twelve countries in Western Europe belong to the European Community (EC): Belgium, Denmark, France, Germany, Greece, Ireland, Italy, Luxembourg, Netherlands, Portugal, Spain and United Kingdom. Austria and Switzerland are members of European Free Trade Association (EFTA). Austria has applied for membership in the EC, and a voting in Switzerland was against the EC. Currently, the European Common Market consists of about 345 million consumers.

At the early stage of the European Market, there was a genuine effort to explore the feasibility of Euro-wheat, Euro-flour, as well as Euro-bread. However, in practice, every country supported its own food specialties, especially in the area of cereal-based food (bread, rolls, confectionery, pasta, breakfast cereals, etc.). So the concept of having Euro-bread never materialized.

However, as cereal-based foods have become increasingly popular in Western Europe, more attention is paid to the quality of the wheat and flours. The increased popularity of cereal-based foods, in part, is due to higher awareness of the population about nutritional benefits of a high-cereal diet and has resulted in remarkable increase in consumption of cereal-based products in most of EC countries, such as Germany (Fig. 4-1). In the European Market, the wheat has to meet minimum requirements (Verordnung EWG 1992) to be sold in the market as food wheat:

- Normal color, sound smell, free of living insects, no sticky doughs
- Broken kernels, max.: 5%
- Shrunken and insect infested kernels, max.: 12%
- Sprouted kernels, max.: 6%;
- Weed seeds and spoiled kernels: max. 3%;
- Falling number, min.: 180 sec;
- Protein content, min.: 9.5% dry basis (d.m.)

Wheat Production and Usage

In Western Europe, nearly all countries overproduce wheat. The wheat classification does not differentiate between hard and soft wheat, but between common wheat and durum wheat.

Table 4-1 shows the production figures for common wheat within the European Community. The total production area is nearly 14 million hectares (ha). The average yields per ha is 5.5 tons. The total production amounts to more than 75 million tons, of which 60 million tons is marketed as food-grade wheat (Commission 1992).

Total wheat usage within the EC is around 45 million tons

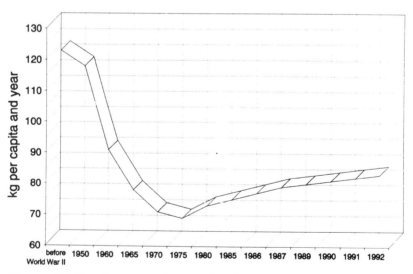

Figure 4-1. Bread consumption trend in Germany.

Table 4-1. Wheat Production and Usage in Different E.C. Countries (1992 Crop)

	Belgium	France	Germany	Greece	Ireland	Italy	Netherlands	Portugal	Spain	United Kingdom	E.C.	Austria	Switzerland
Production													
Area (× 1,000 ha)	211	4,728	2,600	299	100	1,001	127	301	1,732	2,071	13,760	246	102
Yield (dt/ha)	59	65	60	29	70	44	80	8	17	68	55	49	55
Production (× 1,000 t)	1,253	30,837	15,517	858	695	4,376	1,020	253	2,954	14,212	75,550	1,200	570
Usage (USA/E.C.)													
Human	860	4,680	5,040	850	140	5,900	1,000	720	3,350	4,781	27,700	...	...
Animals	267	3,000	2,033	...	340	822	500	140	1,400	3,965	13,420	...	...
Industrial	380	700	475	...	...	60	400	5	30	570	2,620	...	...

Table 4-2. Durum Wheat Production and Usage in Western Europe (1992 Crop)

	France	Germany	Greece	Italy	Portugal	Spain	E.C.	Austria
Production								
Area (× 1,000 ha)	426	14	710	1,601	33	558	3,341	12
Yield (dt/ha)	44	44	24	28	8	22	28	50
Production (× 1,000 t)	1,870	62	1,700	4,524	28	1,210	9,400	60
Usage (× 1,000 t)								
Human	513	250	250	2,143	67	150	3,720	...
Animals	...	...	...	20	...	250	286	...

(including 28 million tons for human consumption and 13 million tons as animal feed). Without export, there would be an overproduction of about 25–30 million tons. The average export figures are now about 20 million tons.

Within the Common Market (Table 4-1), France has the highest wheat production with more than 30 million tons, followed by Germany and the United Kingdom with about 15 million tons each. Netherlands has the highest yield with 8 tons/ha, followed by Ireland and the United Kingdom with about 7 tons/ha. Italy has the highest rate of wheat usage as human food with nearly 6 million tons, followed by Germany (5 million tons), the United Kingdom and France with about 4.7 million tons. United Kingdom and France have the highest rate of wheat usage as animal feed with about 4 and 3 million tons, respectively.

Table 4-2 summarizes the production and usage pattern of durum wheat in Western Europe. The total crop is around 9.4 million tons, and the yield per ha is about 2.8 tons. In the European Market nearly 4 million tons of durum wheat is used as food, especially for production of semolina for pasta production. (Commission 1992; Seibel 1992).

There is a remarkable difference in durum wheat production among various EC countries. Some countries do not grow durum wheat at all (Table 4-2). By far, Italy has the highest durum wheat production with 4.5 million tons, followed by France and Greece with 1.9 and 1.7 million tons, respectively. France and Germany have the highest yields per hectare with 4.4 tons/ha. The yields in Italy, Greece and Spain are much lower.

As shown in Table 4-2, Italy has the highest level of consumption of durum wheat in Western Europe, especially in the form of pasta and noodles. In France more than 0.5 million tons of durum wheat is used for food. In Greece and Germany the amount is about 0.25 million tons each. It is remarkable that in Spain more durum wheat is used for animal feed than for human food.

Wheat production in Austria and Switzerland is relatively low (Table 4-1). Annual production of 1.2 million tons in Austria not only covers the need of the country, but allows for some export to Eastern Europe (Schoeggl 1992). In contrast, Switzerland imports quality wheat (from outside Europe) in order to meet the demand for human consumption and also to improve the quality of the homegrown wheat by blending it with imported wheat. In Switzerland, the total annual usage of wheat for human consumption is about 0.46 million tons, 15% of which is imported (min. 14% d.m. protein).

Quality of Wheat and Flours

In France, the quality of the wheat is based on those flour quality attributes desirable for production of popular baguette-bread. A flour protein level of 11.0–11.5% (d.m.) is necessary. As shown in Table 4-3 for the 1992 crop, the average wheat protein content was 12.5% with a good gluten quality, with no sprout damage (high falling number). Baking strength is mostly determined by the alveograph. A W-value of 211 shows a good baking strength, especially suitable for French bread.

In Germany, there is a much higher demand for flour quality. During the last few years there has been a continuous increase in protein content and sedimentation value of German flour. The improvement on the 1990 to 1992 crops is partly caused by availability of high-quality wheat produced in the eastern parts of Germany (Seibel et al. 1992a; Seibel and Zwingelberg 1992). As shown in Figure 4-2, in the 1992 crop more than 35% of wheat grown in Germany had a protein content higher than 14% d.m. Less than 5% of the wheat crop was lower than 9.5% in protein.

The typical German bakers' flour has an ash content of 0.5%–0.6% d.m. and a protein level from 12.0–2.5%. Figure 4-3 demonstrates that the protein level of German bakers' flour was nearly constant during the last several years with a slight tendency to increase. Based on favorable crop conditions, the falling number has increased since the 1987 crop and addition of malt flour or amylases to the flour has become a common practice. The bread volume has an average level of 650 ml per 100 g flour, using special standardized baking tests (Seibel et al 1992b).

In the Benelux countries (Belgium, the Netherlands, and Luxembourg) wheat protein level ranges between 11 and 12% d.m. These three countries are characterized as "white bread countries," and hence, wheat flours with an ash content of

Table 4-3. Typical Quality Parameters for French Wheat

	Crop 1992
Protein content % d.m. (N × 5.7)	12.5
Sedimentation value, ml	41
Falling number, s	275
Alveogramme (W)	211
Alveogramme (G)	24.1

about 0.6% and a protein level of 12.0–12.5% are needed. Therefore, it is necessary to import high-protein wheat from outside the EC or from Germany.

A similar situation is found in Great Britain (Spencer 1987). The average protein of domestically grown wheat is 11.5–12.0% d.m. For the Chorleywood bread process a protein level of about 12.5% (d.m.) with a high starch damage and high water absorption is necessary (Axford et al 1962; Stewart 1991). In order to obtain such a protein level, high-protein wheat is imported or vital wheat gluten is added to the flour in the mill.

In Spain, wheat protein varies between 10.0 and 15.0%, with sedimentation values between 13 and 50 ml. This wide range in quality is also demonstrated by the W-value of the alveograph, which varies from 80 to 230. Based on the favorable climate conditions, no sprouting damage is experienced (Junta 1992a).

Wheat produced in Austria usually has a very high protein content (1992 crop: 13.5% d.m., wet gluten 33%) (Schöggl 1992). The Austrian bread flour has an ash content of 0.70%

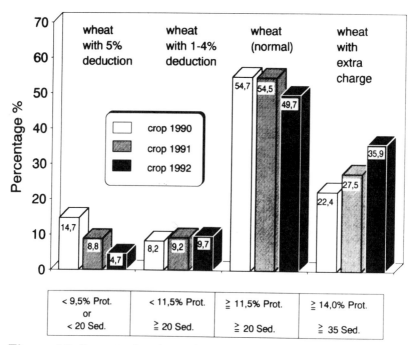

Figure 4-2. Percent of various wheat grades grown in Germany (1990, 1991, and 1992 crops).

d.m. and protein level of 13% d.m. In Switzerland different types of wheat flours are available with an ash content ranging between 0.4 and 0.7% d.m. and a protein level of about 12–13% d.m.

Quality of Durum Wheat

France produces an excellent quality durum wheat with average protein level of 15.6% d.m. and dark, hard and vitreous

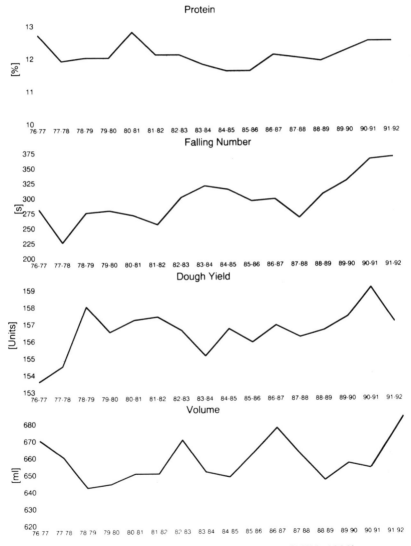

Figure 4-3. Wheat flour quality trend in Germany (1976–1992).

kernels. Sprout damage is experienced in some regions of France (Institut 1992a). Germany produces small quantities of a high-protein content durum wheat with 78% dark, hard and vitreous kernels. The ash content is always a little higher than in France (Seibel 1992; Seibel and Menger 1986). The average protein levels of 1992 crop durum wheat produced in Greece varied between 12.5 and 13.0% d.m. Italy's durum wheat has a low (12–13% d.m.) protein content. The low-protein durum wheat is used for bread production. It is said that durum flour increases shelf life of wheat breads. In Spain the protein level depends on the region. The average is between 13 and 14.5% d.m., with a vitreousness of about 80%. The falling number is usually higher than 300 sec (Junta 1992b). Durum wheat quality is excellent in Austria. Protein content (average: 15.3% d.m.) and vitreousness (average: 90%) are very high, and there is no sprout damage.

Popular Products Made From Wheat

Since EC wheat classification does not assign hard wheat and soft wheat to separate classes, the mills in Western and Central Europe use a wheat blend, consisting of different portions hard and soft wheat, to fulfill the needs of the European bakers. In Europe, bread is made using a lean formula and is perceived to be low in calorie, fat and sugar content. On the other hand, confectionery goods are perceived as high-calorie, based on high amounts of fat and/or sugar. European breakfast cereals mostly consist of whole cereal kernels, and the most important cereal is wheat (besides maize).

Bread and Bread Specialties

Bread consumption, production techniques, and specialties are very diverse in Europe. The ICC (International Society of Cereal Chemistry) Study Group "Nutritional Aspects of Cereal Food" has reported bread consumption in different European countries, as shown in Table 4-4. These figures demonstrate a very high bread consumption in Germany, followed by Italy, Denmark and Portugal. The lowest bread consumption is in United Kingdom and also Luxembourg. The reason for these differences in bread consumption are not well understood, but are mostly based on traditional food habits in the different countries.

France. Official French regulations classify wheat flours according to types, based on ash content. The protein content and protein quality (expressed in France by the W-value of the alveograph) permits the precise classification of flours within the given types (Greenwood et al 1983).

In France, where there is a great selection of bread, about 80% of the bread is still produced in small bakeries. The one traditionally eaten, the baguette, is a white bread with a distinctive appearance (Fig. 4-4). This long, thin, crusty loaf of bread, known throughout the world as French bread and in France as *pain ordinaire*, is actually of Austrian origin, acquired by the French during the Napoleonic wars. French consumers display an insatiable demand for fresh bread, making purchases two or three times each day from small bakeries. This French bread is characterized by a heavy crust (about 40–50%), a crumb with an open grain and an excellent flavor. The bread has a 70-cm length and weighs 250 g. The baguette represents more than 50% of the bread produced in France. The *pain parisien,* the second most popular bread, weighs 400 g, is heavier than the baguette, and has the same length of 70 cm and five lateral cuts (Martin 1992; Tweed 1983).

The "blade cut" made by craftsmen, even in a partially mechanized production, assures that the baguettes retain their original appearance. The recipe for baguette is very simple, with four basic materials: 100 parts flour, 58–66 parts water, 1.8–2.2 parts salt and 2 parts yeast.

The dough is mixed intensively to reach the optimal dough temperature of 25°C.

Fermentation of the dough takes place in three stages. Immediately after mixing, the entire batch is fermented, the batch is then divided into small pieces, and the dough is "knocked back." The final fermentation stage occurs after shaping the

Table 4-4. Bread Consumption in Some European Countries (1991)

Country	kg/person/yr	Population (millions)
Belgium	73	9.8
Denmark	75	5.1
France	56	55.8
Germany	80	80.0
Greece	70	10.1
Ireland	65	3.5
Italy	75	57.2
Luxembourg	52	0.4
Netherlands	64	14.7
Portugal	73	9.8
Spain	70	39.2
United Kingdom	53	57.0

dough into a long cylinder. During this final fermentation the traditional cellular structure of the dough is formed. Baking takes place in an oven heated to approximately 200–250°C in an atmosphere saturated with water vapor. The final proof time varies from 90 minutes to 2 hours.

After being removed from the oven, breads are usually stacked on their ends into loosely woven baskets or open containers to allow free movement of air. Shelf life varies between 4 and 6 hours.

French bread can be produced without any improver, but under French regulations four ingredients may be added that allows the miller or baker to correct possible deficiencies in wheat flour (French National Centre 1977):

- Bean flour with active lipoxigenase to improve the whiteness of the crumb (~2% on wheat flour basis);
- Ascorbic acid to improve the gluten quality (1 g/100kg flour);
- Malt for the correction of the alpha-amylase activity in the wheat flour (~0.3%);
- Lecithin (0.3% of flour weight) to improve gas retention.

Besides the well-known baguettes there are many other French bread specialties, including country breads in the dif-

Figure 4-4. French bread *baguette*.

ferent regions with different shapes. The dough is almost the same as that of the baguettes, but the form differs. Country breads are available as rings or as toasted bread. Very often sourdough and rye flour is added.

The fine wheat bread (*pain du gruau*) is produced from very white flour and may contain milk powder. The standard formula is as follows: 100 parts of very white wheat flour, 65 parts water, 8 parts yeast, 2.1 parts salt, 1.8 parts milk powder and 0.9 parts malt. The duration for the first fermentation is 4 hours and that of the second fermentation 1 hour. Another important French bread is Viennese bread, which is baked from a dough enriched with fat. The average recipe is 100 parts white wheat flour, 60 parts water, 2 parts salt, 1.8 parts milk powder, 1.2 parts malt, 2.5 parts sugar, 6 parts fat, 6 parts yeast and 1 or 2 % baking improver. The three stages of fermentation take about 3 to 4 hours. Viennese bread is generally shaped in the form of baguettes and is usually baked on hot sheets. Its crumb is very fine and distinct from the classic baguette.

The *biscottes* can be compared with the German Zwieback and the Dutch toast. It is a bread recipe with a very high fat and sugar content. The common formula is as follows: 100 parts flour, 5 parts sugar, 5 parts fat, 1 part malt extract, 5 parts yeast, 1.5 parts salt, 50 parts water, and small quantities of ascorbic acid, lecithin and baking improver. In comparison with the production of baguettes and fine wheat bread, fermentation time is shorter; less than 2 hours is required for the different stages.

The view of the French peoples on the quality of their bread varies with their regions and with their habits. But there are critical attributes, not only in France, but wherever French bread is produced:

> French bread should be sufficiently light, so that the crumb is never finely textured. The holes in the crumb should be large in Paris, but in many other regions of France such as in the North, people prefer bread with many holes inside. In the Southern parts of France people like a pale crust. In some parts of the country people like the top of the bread to be floury. In Paris they like a golden crust. But all over the country, "baguette" should be fresh and crunchy (Greenwood et al 1983).

Germany. According to the German food law, baked products are divided into two types: the bread recipe allows a maximum of 10 parts sugar and/or fat to 90 parts of flour; products with more than 10 parts sugar and/or fat to 90 parts of flour in the recipe are called *Feine Backwaren* (confectionery) (Deutsches Lebensmittelbuch 1992; DLG 1992a; Seibel 1991; Seibel et al 1985). In Germany, special criteria for the evaluation of bread and other baked goods including bread specialties were established very early. There are four main groups, depending on the proportion of wheat and rye in the formula. So wheat bread, as flour bread or wholemeal bread, should have at least 90% wheat flour, and rye bread as flour bread or wholemeal bread should have at least 90% rye in the recipe (Meuser et al 1993; Seibel et al 1985).

There are numerous types of breads consumed in Germany. A typical German white bread made from 90% wheat flour is shown in Figure 4-5. During the last few decades, toast-bread has become popular in the market (Fig. 4-6). This bread is baked from wheat flour with the addition of fat, sugar and milk powder. Figure 4-7 shows the most common German bread. This is a mixed grain wheat bread, containing 70% wheat and 30% rye flours. The whole-meal wheat bread, which is widely distributed in the northern parts of Germany, consists of at least 90% wholemeal wheat flours (Fig. 4-8). The difference in loaf volume of the two whole wheat breads shown in Figure 4-8

Figure 4-5. Wheat bread.

Figure 4-6. Wheat toast bread.

Figure 4-7. Wheat mixed bread.

is due to the differences in flour particle size. The loaf on the right is made from a fine milled whole wheat flour, the loaf on the left from a coarse whole wheat meal. Generally, each group is divided into three subgroups, according to the use of the types of milled products: flour bread, meal bread, wholemeal bread.

Figure 4-8. Whole wheat bread.

Figure 4-9. Wheat rolls.

Furthermore, it is possible to prepare a variety of bread specialties by incorporating other ingredients into the formula, for example 10% linseed or germ or milk products. Recently, the use of so-called "non-bread grain," like oat, barley, sorghum, rice or maize has become popular in the bread formula. Each grain may further be specified as flakes, grits, cooked kernels, groats, etc. (Bruemmer et al 1988; Bruemmer and Seibel 1991). A minimum amount of each grain has to be present in the recipe in order to advertise it in the products' label (e.g. linseed bread: minimum 8 kg linseed; milk bread: minimum 50 kg milk; barley bread: minimum 20 kg barley in 100 kg cereal products).

Production of dark breads, including wholemeal flour breads, is on the rise, which has helped to decrease the caloric content per unit and provide a more balanced nutritive composition. Nowadays, even small baked goods (rolls) are made from whole meal products, particularly whole wheat flour. In the past, wheat rolls were only made from white flour. Figure 4-9 is a typical German crusty roll, produced from wheat flour. There are also wheat mixed rolls (70% wheat flour : 30% rye flour) (Fig. 4-10). Nowadays the most popular rolls are made from whole wheat flour (Fig. 4-11). The German consumer has the choice of a wide range of baked goods with different nutritive

Figure 4-10. Wheat mixed rolls.

values. There is a wide range in composition of wheat breads (per 100 g):

- Protein: 8.0–16.3%
- Fat: 3.9–5.2 %
- Carbohydrates: 32.7–51.6 %
- Total dietary fibre: 3.2–9.3 %
- Caloric value: 100–270 kcal

From the total bread consumption in Germany, about 15 % is eaten as rolls. However, due to a remarkable increase in single households, consumption of rolls is on the rise.

The straight dough procedure is mainly used for the production of wheat breads and rolls. All recipe ingredients are added together in one step. Various types of baking improvers are used for the production of bread and rolls to increase loaf volume and improve crumb structure.

A typical wheat roll recipe has 100% wheat flour, 5% yeast, 2% salt, 1% sugar and 2–3% baking improvers. Wheat bread has a basic formula containing 3% yeast, 2% salt, 1% fat, 1% sugar and 2% baking improver. Wheat toast bread is baked with 5% yeast, 2% sugar, 2% milk powder, 3–5% fat and 2% baking improver. For the production of wheat mixed bread (more than 50% wheat flour) 2% yeast and 2% salt is used. 20%

Figure 4-11. Whole wheat rolls.

of the total flour in form of rye flour is used for the production of sourdough. Wholemeal wheat breads/rolls have 2–3% yeast, 2% salt and 2% baking improver in the recipe.

United Kingdom. In the United Kingdom, flour production can be divided as follows (Spencer 1987):

- Bread flours for white, dark and wholemeal bread: 66%
- Biscuit flours: 14%
- Household flours: 5%
- Other uses: 15%

For the well-known Chorleywood bread process (Axford et al 1962; Chamberlain et al 1962) a protein content of 12.5% d.m. is necessary, together with a high starch damage and hence a high water absorption. Very often, vital wheat gluten is added in the mill to reach the necessary protein level.

In the United Kingdom, flour color, instead of ash, is a means to specify flour quality (Kent-Jones and Martin, 1950). A typical bread flour has a color grade of 2.5 and 3.5 units, which is equivalent to a flour yield of 75–77% and an ash content (d.m.) of about 0.55%. Prior to joining the Common Market, about 50% of the annual wheat requirements of UK were imported from Canada and the United States. Since then, the quality of homegrown wheat has improved, and quality wheat is imported from other European countries, especially Germany.

Two main processes are used for production of bread in the United Kingdom: Chorleywood Bread Process (CBP) and Bulk Fermentation. The CBP was introduced in 1961 (Greenwood et al. 1983).

Figure 4-12 demonstrates the most significant differences between Bulk Fermentation and CBP. In the former, basic bread ingredients are mixed in a low-speed mixer. After mixing, the dough is resistant to stretching and has a poor gas retention. Therefore, it is set aside for 3–4 hours to ferment in bulk. In the CBP, dough is fully developed and is processed shortly after mixing to work input of 40 kJ/kg in the presence of an oxidizing agent, compared to 8 kJ/kg with the Bulk Fermentation Process. Most types of bread require baking times between 25 and 45 minutes with a baking temperature of 220 to 250°C. Crusty bread is baked longer than bread which usually has to be sliced. A typical UK loaf is shown in Figure 4-13.

Scotch batch bread is a bread specialty produced by placing dough pieces directly side by side on the oven band. The baking conditions are: 200°C increasing to 300°C, with 70–90 minutes baking time. This bread has a thick top and bottom crust and no side crust. Figure 4-14 shows the three popular breads consumed in the United Kingdom.

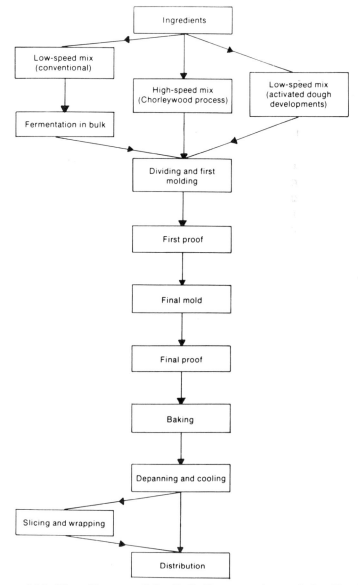

Figure 4-12. Flow diagram of the Bulk Fermentation and the Chorleywood Bread Process.

All U.K. flours must contain at least 0.24 mg of vitamin B, 1.6 mg of nicotinic acid and 1.65 mg of iron per 100 g of flour. Except for wholemeal, wheat meal flour and self-rising flours, all flours are enriched with calcium carbonate between 235 and 390 mg per 100 g of flour.

Wholemeal bread is made from whole wheatmeal and sold generally as a 400-g loaf. Wheat meal bread is made from a dark flour with an extraction rate between 80 and 90%. Wheat germ bread is made from wholemeal (or also white) flour and the addition of 10–25% wheat germ. High-protein bread must contain 22% protein d.m. A high protein level is reached by the addition of vital wheat gluten, milk, and sometimes soy protein. Milk bread is made by addition of 6% whole milk powder to the recipe. Malt bread is a sticky, sweet, dark loaf containing malt and can be made from white wheatmeal or wholemeal flour; it can be plain and also include dried fruit pieces. U.K. rye bread is usually made from a mixture (50:50) of white wheat flour and rye flour.

The U.K. mills produce special cake and biscuit flours from special low-protein wheat varieties that are chlorinated to pH 4.5–5.5. However, chlorination is not allowed under EC rules, and a substitute for chlorination has yet to be identified.

Many of the more educated and young consumers want a daily bread without "chemicals," meaning without additives. In Germany, other preservation technologies have been devel-

Figure 4-13. Typical U.K. loaf.

oped for sliced bread. Only a small amount of sliced bread has microbiological protection by sorbic acid. Nearly all sliced breads are protected against mold by pasteurizing the sliced, wrapped bread (Brack 1990; Brümmer 1990; Seibel 1983, 1984, 1990; Seibel and Ludewig 1983; Verordnung EWG 1991; Verordnung EWG 1993).

Spain. A typical bread recipe consists of 100 parts wheat flour, 1.5–2 parts yeast, 1.5 parts salt, 0.8 parts fat, and 0.5 parts baking improver. Sometimes 10% sourdough is added. Fermentation time varies between 30 min and 3 hours. The bread has a long shape and a weight of 200 g.

In the EC, questions have been raised on the value of treating flour with bromate for bread flour. Most European countries do not allow use of bromate for bread flour. On a volunteer basis, the bromate treatment is replaced by ascorbic acid (addition: 1–3 g per 100 kg white flour).

Cakes and Cookies

France has a wide range of confectionery products. Puff pasties, such as *mille-feuilles* or *chausson*, are types of confections made with a dough consisting of flour, water, salt and shortening (50% of the dough). The dough is laminated 5 times and folded to incorporate all the shortening. Baking takes place at 230–240°C (Greenwood et al 1983).

Figure 4-14. U.K. breads.

Madeleines (Fig. 4-15), also called *boudoirs* or *biscuit à cuillère*, are made from whipped batter, consisting of equal parts of wheat flour, sugar and whole egg. Sometimes the percentage of egg is doubled. *Löffelbiskuit* (Fig. 4-16) is made from the same recipe with a layer of coarse sugar on the surface.

There are two well-known French confectionery goods made from yeasted doughs: The French croissant (originally from Austria) is made with a raised, flaky dough. Its texture is, therefore, the result of steam leavening during baking and the flakiness brought about by the successive rolling of the dough with fat. A typical recipe for croissants is: 100 parts wheat flour, 2.5 parts salt, 6 parts sugar, 5 parts milk powder, 1 part malt, 1 part invert sugar, 3–4 parts yeast and 60 parts water. After 4–5 hours of fermentation, about 55 parts of butter, based on flour weight, is added, and the dough is rolled and folded 3 times; dough pieces are cut, shaped, brushed with egg and baked in the oven (Fig. 4-17).

Brioches are also made from a yeast-raised dough, fermented slowly with straight or sponge fermentation. A typical brioche formulation is: 100 parts wheat flour, 6 parts yeast, 2.5 parts salt, 10 parts sugar, 60 parts eggs, and 50 parts shortening. The dough is mixed to soft consistency, and the shortening is added without excessive mixing. The dough is

Figure 4-15. Madeleines.

permitted to ferment for 4 hours, then cut into pieces. After final 1-hour fermentation the dough is baked. Fermentation imparts superior flavor to the product (Fig. 4-18).

In **Germany**, cakes and cookies are divided into four groups: products from yeasted doughs (e.g. *Weihnachtsstollen*, rusk), products from non-yeasted doughs (e.g. biscuits, ginger bread), products from whipped batters (e.g. sponge cake), and products made from unwhipped batters (wafers) (DLG 1992a; Seibel 1991; Seibel et al 1985). A typical yeast-raised product

Figure 4-16. *Löffelbiskuit.*

Figure 4-17. Croissants.

is *Stollen*, traditionally baked for Christmas. Stollen is usually baked by the sponge dough method. The sponge consists of wheat flour, yeast, water and sugar, mixed to a temperature of about 28°C and allowed to rest 20 minutes. A pre-mixture is prepared from wheat flour, margarine, sugar, marzipan, citrus fruit peel, salt and water. This pre-mixture with a temperature of 28°C is also allowed to rest for 20 minutes. Then the sponge and the pre-mixture is mixed together with raisins, sugar and other ingredients. After a resting time of 20–30 min, baking takes place for 35–40 min, decreasing the temperature from 230°C to 190°C. Steam is added at the beginning of the baking. After cooling for 1 hour, margarine is spread on the product and later covered with a special sugar. According to the official description, Stollen should have at least 30% fat

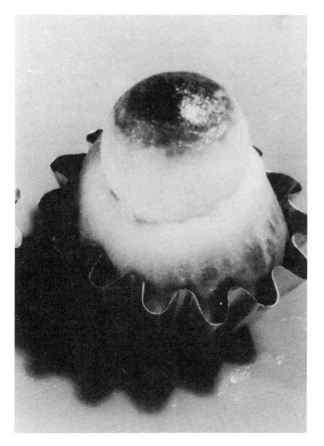

Figure 4-18. Brioches.

(margarine or butter) and also 60% raisins, flour basis (Fig. 4-19).

Biscuits are baked from non-yeasted doughs. The basic formulation for biscuits includes wheat flour, margarine or butter (minimum 10%), sugar, water, and salt. The dough is mixed for 25–30 minutes at low speed to a dough temperature of 40°C. The dough is rolled several times and laminated. After repeated rolling, it is cut into appropriate pieces. The baking conditions (with little steam) are 8–9 min at 200–220°C.

The typical German cake is *Sandkuchen* (Fig. 4-20), made from a whipped batter containing at least 20% each of egg, butter or margarine, calculated on batter weight. The batter is

Figure 4-19. *Stollen*.

Figure 4-20. *Sandkuchen*.

produced usually according to the all-in procedure (see Chapter 1). The batter is filled into pans and baked for 45 min at 190–200°C.

United Kingdom. The number of U.K. cakes and biscuits is very large. Cakes are often distinct to certain regions, with

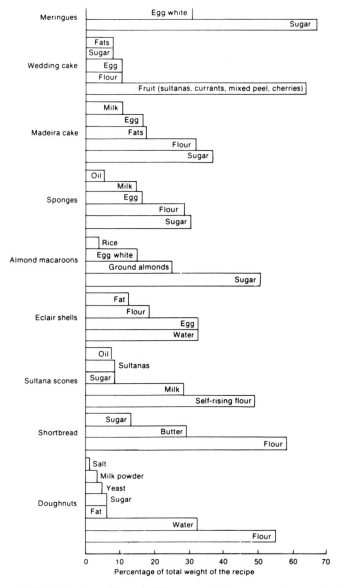

Figure 4-21. Main ingredients of flour confectionery recipes made in the United Kingdom. .

large variations. Figure 4-21 presents typical recipes of confectionery products in the U.K. Strong flour is used for yeast-fermented products such as doughnuts, buns, Danish pastry and puff pastry. Soft flour is used for fruit cakes, cones, short-cakes, and so on. Chlorinated soft flour is used in sponge and Madeira cakes, Swiss rolls and angel cake. Many ingredients are included in confectionery recipes: eggs, milk and milk products, different types of fat, emulsifiers and stabilizing agents. The light cellular structure of many confectionery products is achieved with aeration by mechanical, chemical or fermentation techniques (Greenwood et al 1983).

The principal ingredients of biscuits are flour, fat and sugar. The finished biscuits have a low residual moisture content of about 1 to 4%.

The main U.K. biscuits are:

- Cream crackers and butter puffs, made from hard doughs
- Semisweet biscuits (e.g. Marie, Osborne, *Petit beurre*), made from hard sweet doughs
- Shortbread, short cake and short biscuits, made from short doughs with high fat and sugar levels

Wafers are made from very fluid batter, which consists of flour, water, vegetable oil, salt, lecithin and sodium bicarbonate. The baking is done between two special baking hot plates, and the baked wafer sheet is ejected by compressed air. A wide range of finish products is produced on the basis of wafers.

Breakfast Cereals

For the last decade, ready-to-eat (RTE) breakfast cereals and other non-baked cereal products have become increasingly popular, partly due to new dietary guidelines encouraging consumers to consume more cereal foods, especially wholemeal products. There are a large number of new breakfast cereals on the market that can be divided in different groups (DLG 1992b; Seibel and Nestl 1988):

- Whole kernels of wheat (normal wheat, durum wheat, spelt wheat), triticale, rye, barley, oat, maize, rice, sorghum, buckwheat
- Ground cereals
- Flaked cereals
- Puffed cereals
- Extruded cereals
- Longitudinal or cross-cut kernels

Muesli is a RTE cereal which consists of cereals, oilseeds and dried fruits. A new line of hard and soft breakfast bars with large amounts of specially treated wheat kernels is becoming popular with European consumers.

Popular Products Made from Durum Wheat

Pasta products currently enjoy a slowly increasing popularity with the consumer. A typical pasta-eating country is Italy, with a consumption 4–5 times higher than in Germany. France is somewhere in between. Besides Greece, Portugal and Spain, the other Western European countries have a low pasta consumption.

Pasta product types can be divided as follows (Pavan 1991):

- Pasta products, dried, usually made from durum semolina
- Pasta products, dried, usually made from durum semolina and soft wheat semolina/flour
- Pasta products, dried, usually made from soft wheat semolina/flour
- Egg-based pasta, dried, usually made from durum semolina or a mixture of durum semolina with soft wheat semolina/flour or soft wheat semolina flour, with egg added
- Wholemeal pasta products, dried, usually made from durum wheat wholemeal semolina or a mixture of durum wheat wholemeal semolina and soft wheat wholemeal semolina/flour or soft wheat wholemeal semolina/flour, with or without egg added
- Pasta products, altered in its nutritional value, dried (e.g. enriched with protein or dietary fiber)
- Pasta products, dried, made from non-bread cereal grains (e.g. corn, buckwheat, millet)
- Dietary pasta products, dried, made from different kinds of cereal grains (e.g. gluten-free pasta, made without wheat, rye, oat and barley), especially for the nutrition of people with celiac disease

In some European countries, such as Greece, Italy and France, only durum wheat is used for pasta production (Resmini 1968). Some pasta products contain eggs at the level of 3 eggs/1 kg semolina. The pasta cooking quality is optimal even with using up to 40% soft wheat semolina. In Germany, a semolina blend for the production of pasta consists of 66% durum semolina and 34% soft wheat semolina/flour. Spaghetti is still produced from 100% durum, and noodles for soups, on the other hand, are made from 100% soft wheat semolina or flour (Seibel 1992).

"Fresh" pasta with a shelf life of 2–3 days if refrigerated or weeks and months if frozen at –18°C is becoming very popular, partly due to the better quality of fresh pasta foods directly compared with home preparation of fresh pasta. In some European countries, fresh pasta products are divided into three general groups (Pavan 1991):

• Unpacked fresh pasta products; shelf life: 3–4 days
• Packed fresh pasta products; shelf life up to 40 days
• Packed fresh pasta products; shelf life up to 120 days

All fresh pasta products have relatively high moisture contents (~30%) and hence, are prone to spoilage. Therefore, strict microbiological quality control during processing, packaging and storage is needed. By pasteurization with steam or immersion in hot water, the fresh pasta can be brought to temperatures of at least 85°C within a short time. As a result, the possibility of the bacterial spoilage is minimized. Also the formation of mycotoxins is avoided. Use of hot air at 95–97°C is another means of pasteurization.

Consumer and Technological Trends in Wheat Usage

Generally, consumers perceive cereal-based foods positively and consider them to be healthful. But there are also consumer groups who believe that wheat contains heavy metal residues (cadmium, lead) and pesticides. Therefore, wheat production in Europe is divided in three different sections (Bolling et al 1986; Seibel 1990; Verordnung 1991, 1993):

• Organic production
• Controlled production
• Integrated production (conventional production)

The organic production is the former biological production that is well defined by recent EC procedures. Use of fertilizers and plant protection substances is prohibited. The process needs official certification starting with the cultivation of organic wheat to the production of organic bread in the bakery. All production parameters are controlled by independent institutes.

The integrated wheat production uses all the knowledge

about plant nutrition and plant protection to achieve an optimal result in wheat quantity and also wheat quality. Much progress was made in breeding disease-resistant cultivars, and so the use of plant protection agents has decreased during the last few years. The use of fertilizers is also optimized by better detection of nutrients in the soil.

The percentage of nutritionally aware consumers is rapidly increasing, and consequently, the consumption of wholemeal wheat products will increase. To date, consumption of bread and rolls made from whole wheat meal has increased moderately. On the other hand, breakfast cereals and other cereal-based foods have increasing rates of more than 10% per year during the last few years.

The International Association of Cereal Science and Technology (ICC) Study Group "Nutritional Aspects of Cereal Foods" has studied the different guidelines from European countries and worked on a recommendation which summarizes the future of cereal-based foods and also wheat foods in an excellent way. In its international nutritional recommendations and implications for consumption of cereal foods, it is recommended that intake of the following nutrients be increased:

- Complex carbohydrates (polysaccharides)
- Dietary fiber
- Plant proteins

On the other hand, it is recommended that intake of these nutrients be reduced:

- Simple carbohydrates
- Fat (especially saturated fat)
- Cholesterol

In addition to these recommendations regarding nutrient intake, direct reference is often made to foods. In such cases, increased consumption of the following foods is recommended:

- Cereal foods/bread (especially whole grain products)
- Fruits and vegetables
- Fish

Intake of these foods should be reduced:

- Sweets
- Fatty foods
- Meat products
- Alcoholic beverages

As a result of their nutrient compositions, cereal foods and breads are particularly well-suited for diets that follow the above recommendations: starchy carbohydrates with high molecular weights are among the most quantitatively significant nutrients in cereal foods. With their high dietary fiber-content, cereal foods and baked goods are an important source of such fibers—often the most important one. In addition, they supply considerable amounts of plant proteins, and only contain small quantities of low-molecular-weight carbohydrates, fat and cholesterol, etc. The ICC's study group on nutrition has concluded that "average daily consumption of cereals should range between 250 and 350g per person. In addition, preference should be given to cereal foods and baked goods containing the largest possible portions of whole grain products." Regarding technological trends, it is remarkable that in many bread recipes more and more other cereals, such as barley, oats, maize, rice, sorghum and buckwheat, are being used. There is also a trend to use only baking improvers which are really necessary and at the optimal dosage. In Germany, for example, nearly all bakers changed from acidification agents to natural sourdough for the production of rye and rye mixed bread (Bruemmer 1990; Seibel and Bruemmer 1991). It was already mentioned that mold protection is only accomplished by pasteurization and not by adding sorbic acid.

Today, many consumers select food for the "freshness" aspect of it. Therefore, the sale of frozen dough has increased during the last five years in order to shorten the time between production of baked products in the bakery and consumption of these products at home. There was a trend during the last decades to shorten fermentation times by increasing fermentation temperature and to shorten baking times by using different baking systems. But all these experiments were not really successful. For a bread with excellent quality and taste we still need optimum fermentation and minimum baking times.

References

Autran, J.C. 1989. Soft wheat: View from France. Cereal Food Worlds 34: 667.

Axford, D.W.E., Chamberlain, N., Collins, T.H. and Elton, G.A.H. 1962. The Chorleywood bread process. Cereal Sci. Today 8:265.

Bolling, H., Gerstenkorn, P. and Weipert, D. 1986. Vergleichende Untersuchungen zur Verarbeitungsqualität von alternativ und konventionell angebautem Brotgetreide. Getreide Mehl u. Brot 40:86.

Brack, G. 1990. Herstellung von alternativen Feinen Backwaren. In: Seibel, W. (ed.) Bio-Lebensmittel aus Getreide. Hamburg: Behr's Verlag p. 53-66.

Bruemmer, J.-M. 1990. Herstellung von alternativem Brot und Kleingebäck. In: Seibel, W. (ed.) Bio-Lebensmittel aus Getreide. Hamburg: Behr's Verlag p. 31-51.

Bruemmer, J.-M. 1991. Modern equipment for sourdough production. Cereal Foods World 36:305.

Bruemmer, M.-M., Morgenstern, G. and Neumann, H. 1988. Herstellung von Hafer-, Gerste-, Mais-, Reis-, Hirse- und Buchweizen-Brot. Getreide Mehl u. Brot 42:153.

Bruemmer, J.-M. and Seibel, W. 1991. Spezialbrote mit Nicht-Brotgetreidearten und/oder Ölsamen. Getreide Mehl u. Brot 45:45.

Chamberlain, N., Collins, T.H., and Elton, G.A.H. 1962. The Chorleywood bread process. Baker's Dig. 36:52.

Commission of the European Community. 1992. Bilans prévisionnels. Campagne 1992/92.

Deutsche Landwirtschafts-Gesellschaft (DLG) (ed.) 1992a. Prüfbestimmungen für Brot und Feine Backwaren. Frankfurt/M.: DLG-Verlag.

Deutsche Landwirtschafts-Gesellschaft (DLG) (ed.). 1992b. Prüfbestimmungen für Getreidenährmittel. Frankfurt/M.: DLG-Verlag.

Deutsches Lebensmittlebuch 1992. Leitsätze für Feine Backwaren. Bundesanzeiger 44/86b:5.

Didone, G., and Pollini, C.M. 1991. Kontrolle der Maillard-Reaktion im THT-Trocknungsverfahren. Getreide Mehl u. Brot 45:216

Douglass, J.S. and Matthews, R.M. 1982. Nutrient content of pasta products. Cereal Foods World 27:558.

French National Center for Coordination of Studies and Research on Nutrition and Food. 1977. Recueille des usages concernant les pains en France. Colloque "Le Pain" et journées de information 15./15.Nov.

Gavin, M. 1993. Haltbarkeit von Teigwaren, Instant-Teigwaren und Couscous. Getreide Mehl u. Brot 47: in press

Gough, G.M., Whitehouse, M.E. and Greenwood, C.T. 1978. The role and function of chlorine in the preparation of high-ratio cake flour. Crit. Rev. Food Sci. Nutrition 10:91.

Greenwood, C.T. Guinet, R. and Seibel, W. 1983. Soft wheat uses in Europe. In: Yamazaki, W. and Greenwood, C.T. (eds.) Soft wheat. Production, Milling, Breeding and Using. AACC, St.Paul/Minnesota, p. 209-266

Institut Technique des Cereales et des Fourrages (ITCF) and Office National Interprofessionnel des Cereales (ONIC). 1992a. Qualité de blés dur, récolte de France 1992.

Institut Technique des Cereales et des Fourrages (ITCF) and Office National Interprofessionnel des Cereales (ONIC). 1992b. Qualité des variétés des blés tendres, récolte de France 1992.

Junta de Andalucia. 1992a. Encuesta de Calidad de los Trigos Blandos Españoles. Cosecha 1992.

Junta de Andalucia. 1992b. Encuesta de Calidad de los Trigos Duros Españoles. Cosecha 1992.

Kent-Jones, D. W., and Martin, W. A. 1950. A photoelectric method of determining the colour of flour as affected by grade, by measurements of reflecting power. Analyst 75:127-133.

Link, P. 1979. Cerealien - Definition und Marktübersicht. Getreide Mehl u. Brot 33:327.

Manser, J. 1981. Optimale Parameter für die Teigwarenherstellung am Beispiel von Langwaren. Getreide Mehl u. Brot 35:75.

Manser, J. 1986. Einfluß von Trocknungs-Höchsttemperaturen auf die Teigwarenqualität. Getreide, Mehl u. Brot 40:309.

Martin, G. 1992. Des exigences précises et deversifiées (La panification en France). Perspectives Agricoles 174:20.

Meuser, F., Bruemmer, J.-M., and Seibel, W. 1993. Bread varieties in the middle of Europe. Cereal Foods World (in print).

Ocker, H.D., and Brueggenmann J. 1986. Schwermetalle und Pflanzenschutzmittelrückstände bei Getreidenährmitteln. Getreide Mehl u. Brot 40:12.

Pavan, G. 1991. Markttendenzen und technologische Aspekte der Frischteigwaren. Getreide Mehl u. Brot 45:156.

Resmini, P. 1968. Produzione di pasta. El'Uovo l'industria pasteria 73:4.

Schoeggl, G. 1992. Getreideernte 1992 - Qualität und Auswirkungen auf die Mehlbeschaffenheit. Mühlenmarkt 93:348.

Seibel, W. 1983. Alternative Backwaren, Herstellung und Qualität. Getreide Mehl u. Brot 37:7.

Seibel, W. 1984. Ernährungsphysiologische Beurteilung normaler und alternativer Backwaren. Bäcker-Ztg. 22:10.

Seibel, W. (ed.). 1990. Bio-Lebensmittel aus Getreide. Hamburg: Behr's Verlag.

Seibel, W. (ed.). 1991. Feine Backwaren. Berlin/Hamburg: Parey.

Seibel, W. 1992. Internationale Durumweizensituation in den Getreidewirtschaftsjahren 1990/91 und 1991/92. Getreide Mehl u. Brot 46:227.

Seibel, W., and Bruemmer, J.-M. 1991. The sourdough process for bread in Germany. Cereal Foods World 36:299.

Seibel, W., Bruemmer, J.-M., Menger, A., Ludewig, H.-G. and Brack, G. 1985. Brot und Feine Backwaren. Arbeiten der DLG Band 185. Frankfurt a.M., DLG-Verlag

Seibel, W., Gerstenkorn, P., and Klotz, D. 1992a. Die Qualität der deutschen Weizenernte 1992, 1. Teil: Quantitatives und qualitatives Ergebnis in Bund und Ländern. Mühle u. Mischfuttertechn. 129:579

Seibel, W., and Ludewig, H.-G. 1983. Begriffsbestimmung und Herstellung alternativer Backwaren. Brot u. Backwaren 31:13.

Seibel, W., and Menger, A. 1986. Ergebnisse mehrjähriger Sortenprüfung bei Durumweizen. Getreide Mehl u. Brot 40:261.

Seibel, W., and Nestl, B. 1988. DLG-Qualitätsprüfung für Getreidenährmittel (Systematik - Begriffsbestimmungen - Prüfschema). Getreide Mehl u. Brot 42:21.

Seibel, W., and Zwingelberg, H. 1992. Die Qualität der deutschen

Weizenernte 1992, 2. Teil: Mahl- und Backqualität von Weizensorten und -partien aus der Bundesrepublik Deutschland und importiertem französischem Weizen. Mühle u. Mischfuttertechn. 129:612.

Seibel, W., Bruemmer J.-M., and Morgenstern, G. 1992b. Qualität der Roggen- und Weizenmahlerzeugnisse im Getreidewirtschaftsjahr 1990/91. Getreide Mehl u. Brot 46:108.

Spencer, B. 1987. Zur Situation der Mühlen in Großbritannien. Getreide Mehl u. Brot 41:198.

Stewart, B.A. 1991. The 1990 wheat harvest. Bull. Flour Milling and Baking Res. Assoc. February 3.

Stroh, R. 1991. Mikrobiologische Beschaffenheit verschiedener Frischteigwaren aus dem Handel. Getreide Mehl u. Brot 45:186.

Tweed, A.R. 1983. A look at French bread. Cereal Foods World 28:397

Verordnung (EWG) Nr. 2092/91 des Rates vom 24. Juni 1991 über den ökologischen Landbau und die entsprechende Kennzeichnung der landwirtschaftlichen Erzeugnisse und Lebensmittel. Amtsblatt der Europäischen Gemeinschaften nr. L198/1 vom 22.07.1991.

Verordnung (EWG) Nr. 689/92 der Kommission vom 19. März 1992 über das Verfahren und die Bedingungen für die Übernahme von Getreide durch die Interventionsstellen. Amtsblatt der Europäischen Gemeinschaften Nr. L74 vom 20.03.1992.

Verordnung (EWG) Nr. 207/93 der Kommission vom 29. Januar 1993 zur Festlegung des Inhalts des Anhangs VI der Verordnung EWG Nr. 2092/91. Amtsblatt der Europäischen Gemeinschaften Nr. L25/5 vom 02.02.1993.

Wade, P. 1972a. Flour properties and manufacture of semi-sweet biscuits. J. Sci. Food Agric. 23:737.

Wade, P. 1972b. Technology of biscuit manufacture: Investigation of the process for making cream crackers. J. Sci. Food Agric. 23:1213.

CHAPTER 5

Wheat Usage
in Eastern Europe

Radomir Lásztity

Albania is a small country in the Balkan peninsula with an area of 28,000 km and a population of 3.2 million. It has been long isolated, not only from Western Europe but also from Eastern Europe. Although political changes in Eastern Europe have reached Albania, and a transition period to democracy and market economy is apparently underway, knowledge of the details of the production, processing and use of cereals is still poor.

Roughly 20% of the country's land is given over to agriculture. The main agricultural products, maize, wheat and beans, play an important role in meeting Albania's food consumption. Data on cereal and specifically wheat production by Albania is summarized in Table 5-1. Because its wheat production does not cover the domestic consumption, Albania is a wheat-importing country. Bread (either wheat flour or mixed wheat and maize flour) is the main food product made with wheat.

Bulgaria. Although wheat and maize are the main cereals grown in Bulgaria, considerable amounts of barley are also produced. Rye and oats are also grown, but in small quantities. Bulgarian cereal and wheat production is summarized in Table 5-1. Most Bulgarian wheat cultivars are soft red winter types with widely differing baking qualities. Efforts made to breed high-yielding soft wheat cultivars adapted to Bulgarian climatic and soil growth conditions (Popov et al. 1969) relied primarily on Italian cultivars (e.g., San Pastore). Smaller amounts of hard wheat and durum wheat cultivars are also grown. Practically all the wheat produced is used for food.

It is somewhat difficult to differentiate products made from soft or hard wheat, except for pasta. Bread is the main wheat-based food product in Bulgaria, and its consumption (~150 kg/capita/year) is high in comparison to that in Western Europe (Chapter 4). In the pre-World War II period, most of the bread was baked at home both by farmers and by many city dwellers. Because of rapid industrialization after the World War, that situation changed, and most of bread production was done by large state or cooperative bakeries. However, political and economic changes of recent years have resulted in the privatization of part of the state-owned industry; consequently, many smaller, private bakeries are now open.

More than 30 types of breads and more than 100 types of breakfast baked products are produced commercially in Bulgaria. The production of macaroni products is relatively low. *Sofia* bread, one of the most popular types, is baked from whole wheat flour by a sponge-and-dough procedure. A starter sour, made at the first stage, is used to make the main dough. The sour is high in water (100 kg flour and 80 kg of water) and contains all the yeast. Sponge, once mixed, is fermented for 6–8 hours at a setting temperature of 27–29°C, mixed with an additional 5% flour, and fermented for another 40 minutes. The final dough is prepared by mixing the starter sour with the rest of flour and 1.4–1.7% salt (on flour basis) and about 15% dough from a previous bake (to increase acidity). This dough is fermented for 30–40 minutes at 26–29°C, remixed and fermented for another 20–30 minutes.

The dough is scaled, rounded, allowed to rest 5 minutes, shaped in the oblong loaves, placed on long wooden pallets, and given a final proof of 30–35 minutes at 30–32°C at a relative humidity of about 85%. Baking is done for 40 minutes at 220–230°C. The final product has a high volume but a coarse crumb texture.

Two other popular bread types, *Stara Zagora* and *Dobrud-*

Table 5-1. Cereal and Wheat Production in Albania and Bulgaria (1,000 tons)

Year	Albania		Bulgaria	
	Cereals	Wheat	Cereals	Wheat
1983	1,067	583	7,932	3,608
1984	1,042	580	9,251	4,836
1985	1,003	550	5,384	3,068
1986	997	555	8,492	4,327
1987	1,010	565	7,284	4,141

Source: FAO Statistical Yearbook, 1990

ska, are also baked by this type of sour/sponge procedure but from white wheat flour.

Various types of breads, rolls and other bakery products in Bulgaria differ widely in size, shape, fillings and coatings. Many contain ingredients such as jams, jellies, cheese and may be sprinkled with poppy, sesame, salt, sugar and various spice seeds.

Among the more popular is *kolatsch,* a disc-shaped sweet bread. It is commonly part of festive or celebration meals. The bread is usually ornamented and decorated, the motifs of which are taken from agriculture and religion.

A typical Bulgarian savory, dough-based product is *Banitza,* made from thin layers of folded and unfermented dough filled with dairy products, meat, fruits, and vegetables.

The thin dough, which contains white wheat flour, water, salt, fat, sunflower seed oil and eggs, is allowed to stand at 19–21°C for 15–20 minutes after mixing. After resting, the dough is formed into small round pieces by hand or by machine and allowed to rest for 30–40 minutes, then sheeted into 2 to 3 mm thin layers of 25–30 cm in diameter. Up to four layers are placed, one on top of the other, and the sheet is worked into almost transparent layers (0.2 mm). The filling is placed in the center, the dough edges sealed with egg wash, folded, then baked at 250–275°C on flat pans. A similar type of product is known as *milinki.*

Mekizi is a fried product made from layers of yeast-leavened wheat flour dough. Slack dough (1 kg flour, 0.7 kg water, yeast and salt) is fermented for 3–4 hours at 26–27°C, scaled into 36–38 pieces, punched, proofed for 30–40 minutes, and fried in sunflower oil for several minutes at 220–230°C. The final product is round or elliptical with a thin center and thick outer rim. The fried product is generally sprinkled with powdered sugar and served hot. Another Bulgarian specialty is the sesame-covered *gevretzi.*

Czech and Slovak Republics. Similar to the general European trend, wheat production in Czechoslovakia increased significantly in the last decades. The main wheat growing regions are north-west Bohemia, southern Moravia and south-west Slovakia. This increase is connected, similarly, with higher-yielding cultivars. Yearly production ranged in the last decade from 4 to 6 million tons.

Most of the wheats grown in Czechoslovakia are of intermediate baking quality. The amount of high-quality wheat grown is relatively small and is mainly soft wheat. Durum wheat is not grown in commercial amounts but is grown at several experimental stations. Approximately 1.5–2.0 million tons of wheat are used for food purposes. The remainder of the crop serves as feed and seed. Foreign trade in wheat is not significant. The data summarized in Table 5-2 were reported prior to disintegration of the country to Czech and Slovak Republics.

A large number of soft wheat flour-based products are produced and consumed in Czechoslovakia. Cookies are produced from chemically leavened, high-fat, high-sugar doughs. Saltine crackers and cream crackers, both yeast-leavened and chemically leavened are produced in Czechoslovakia. Of the wafers produced in Czechoslovakia, the best known is *Karlovarske oplatky* (*Karlsbader Oplatten* in German), round shaped plates filled with a mixture of finely ground sugar, high melting point fat, ground nuts, flavorings and finely ground scrap of wafers.

Perniky (medovnik) is a honey cake similar to those prepared in many European countries. Commercial *perniky* is produced from white soft wheat flour (although some traditional types may contain rye flour), invert sugar syrup and/or fructose syrup, honey, glucose syrup with addition of some fat (edible oil), eggs, ground oilseeds, chemical leavening agents, flavor and coloring agents. The dough, prepared by mixing the flour and invert sugar, has a long rest time lasting from 3–4 days up to months. After resting, the leavener and remainder of ingredients are added to the dough, the dough divided, and then formed into a tremendous variety of shapes (figures) and sizes. Baking is at 230–300°C for 8–12 minutes. After cooling, *perniki* is often iced and/or decorated.

Foam cakes, as well as specialty sweet yeast-raised goods are also known and produced in the country (Fig. 5-1). A Slo-

Table 5-2. Wheat Use in Czechoslovakia[a,b]

Year	Use (100 tons)		
	Food	Seed	Feed
1986/87	1,944.8	251.9	3,318.9
1987/88	1,947.8	271.6	3,988.3
1988/89	1,963.3	281.2	4,325.7
1989/90	1,873.9	320.6	4,292.1
1990/91	1,537.5	325.6	4,441.5
1991/92	1,573.1	233.7	3,715.0
1992/93[c]	1,602.0	227.9	3,765.5

[a] 1,000 tons
[b] Projection
[c] This data was reported prior to disintegration of the country to Czech and Slovak republics.

vakian specialty is the filled roll named *Zemplinsky kolac* (Zemplin cake).

In Czechoslovakia, as in other Eastern European countries, bread is the main food product produced from wheat. Mixed wheat-rye bread is the most popular, accounting for about half of the total consumption. Wheat bread and rolls rank second and rye bread third. According to Sabo and Kozel 1983), the annual per capita consumption of milled and/or bakery products in Slovakia was as follows: bread, 66 kg; bread rolls, 22 kg; pastry, 5.4 kg; cereal-based confectionery products, 2.3 kg; and flour for home uses, 21.2 kg.

More than 20 types of breads are produced in Czechoslovakia, the two main types of mixed flour bread are the "choice" bread and the "common" bread. The former is produced from medium-graded wheat/rye flour (70:30). Because of this, the bread has a wheat bread character. A sponge-and-dough procedure is commonly used to make the dough. The bread is ob-

Figure 5-1. Fine baked goods in Czechoslovakia.

long and hearth-baked and occasionally panned. In Slovakia, where this bread is most popular, addition of 20% buttermilk and 0.5% potato flour to the formula is a common variation. On the farms, mixed breads are baked with the addition of caraway seeds.

"Common" bread, also known as brown bread, is made from lower-grade rye and wheat flours and contains a higher proportion of rye flour (60–80% of total flour). Breads with a high proportion of rye flour are sour leavened.

All wheat flour breads, known generally as *white bread*, are produced in a wide variety of sizes and shapes, mainly by a sponge-and-dough procedure. A number of whole-grain breads are also common (e.g. Graham, Kiev and Moscow type breads). A particular specialty sour-type bread is made from 50% white rye flour, 32% whole ground rye, and 18% wheat flour of medium extraction.

All the common types of pasta (macaroni products) are produced in Czechoslovakia. However, they are produced mainly from hard winter wheat farina, rather than durum semolina.

Among the other wheat flour-based products are cereal-based baby foods and frozen dough products. A Czech specialty soft wheat product is the yeast-leavened cookies *knedlik*.

Hungary. Climatic conditions in Hungary, along with its soil and its terrain, provide favorable growth conditions for cereals. The main agricultural products of Hungary are wheat and maize. Some barley is grown at relatively higher elevations, but the production of rye, oats and triticale is low.

As shown in Table 5-3, a slight decrease in area and production has characterized Hungarian wheat production over the past decade. In the last two years, production dropped from 5 million to 3.5 million tons/year, principally due to the loss of Eastern markets and consequent difficulties in reorganization

Table 5-3. Wheat Production in Hungary

Year	Area Harvest (1,000 ha)	Production (1,000 tons)	Yield (kg/ha)
1982/83	1,313	5,761	4,390
1983/84	1,355	5,985	4,415
1984/85	1,361	7,367	5,420
1985/86	1,356	6,555	4,830
1986/87	1,318	5,744	4,360
1987/88	1,288	7,026	5,450
1988/89	1,248	6,540	5,240
1989/90	1,190	5,420	4,647
1990/91	1,035	4,980	4,810
1991/92	869	3,600	4,185

of the agricultural system. Only winter wheat is grown in Hungary. The yield of spring wheats is always questionable, due to the frequent shortage of rain in spring time. Until the 1950s, the majority of these wheats were hard-winter-red types with very good baking strength. Practically the total wheat crop was used for food purposes. However, since the early 1960s, a steady increase in the growth of new, high-yielding cultivars from the former Soviet Union (Bezostaya and Julileynaya), Italy (San Pastore) and Yugoslavia (Partizanka) has occurred. These new cultivars had a wider range of baking strengths. The increased production (6–7 million tons/year in the late 1980s) not only satisfied food demands of 1.5–2 million tons/year, but also allowed a major increase in wheat export as well as usage in the production of formulated feeds (blended with maize).

Currently, Hungarian standards distinguish two types of wheat: food or bread wheats and feed wheat. The first type consists mainly of hard winter cultivars with good baking strength. They are used for export purposes and the production of baked goods and pasta. The second group contains soft and intermediate cultivars and serves for production of formula feeds.

Durum wheat is grown only in small quantities. In Hungary, the traditional pasta products are produced from high-quality hard winter wheat farina. In recent years, efforts have been made to increase durum wheat production with the introduction of several new cultivars. The future of durum wheat growth is dependent on the possibility for export of durum-based Hungarian pasta products.

To date, very few high-quality soft wheat cultivars are grown in Hungary. This makes it difficult for the cake and biscuit industry to identify and procure good raw materials for manufacture of traditionally soft wheat based products.

The analysis and control of wheat flour baking quality and the development of dough testing methods has been important in Hungary for many years. In fact, such well known methods as the Pekar test, farinograph and laborograph were invented and first used in Hungary. Today, wet gluten content and quality and farinograph curve analysis are the most commonly used quality control tests for wheat and flour. While the falling number test is often used, extensograph or alveograph use is not common.

The industrial use of the wheat for the production of starch and vital gluten has grown significantly in recent years. However, it still represents a small percentage of the total wheat usage in any given year.

As a country, Hungary typically consumes a large amount of baked goods. Consequently, wheat bread and wheat-flour based fine bakery products form the overwhelming majority of the baked products. A great variety of soft wheat flour products are produced in Hungary, both yeast and chemically leavened. Cookies, foam cakes, biscuits, pie crusts, crackers, pretzels and wafers are the best-known products.

The wheat breads produced commercially by large bakeries weigh from 0.5 to 2.0 kg. In general, they are round or oblong, and almost always hearth-baked. Typically, after proofing and just before baking, they are scored on the top surface. The final products are characterized by high volume and a porous and elastic crumb structure.

The technology used to produce most Hungarian breads is a two-step (sponge-and-dough) procedure which employs yeast at low levels and relatively short dough fermentation times. The resulting product is aromatic, with a mild sour taste. Use of flour improvers is not common. Only ascorbic acid is permitted according to Hungarian food regulations. However, emulsifiers, mainly monoglycerides, phosphatides and calcium stearoyl/palmitoil/lactylates are widely used.

Three types of wheat flours (or mixtures of these types) are used for Hungarian breads; type 550 (fine white), type 800 (white) and type 1120 (extra white), depending on the type of bread. A typical formula (calculated on 100 kg of bread) for white wheat bread is as follows:

wheat flour (type 800)	74.0 kg
yeast (0.4%)	0.3 kg
salt (2.0%)	1.5 kg
water (58–60%)	43.0–44.5 kg
(ascorbic acid 2 g/100 kg flour)	
(emulsifier 0.1–0.2%)	

Sponge formulation and processing, again calculated on 100 kg bread, are as follows:

wheat flour (type 800)	26–37 kg
yeast	0.3 kg
water	27–30 kg
initial temperature	26–28°C
final temperature	28–30°C
fermentation time	5–6 hours

After sponge fermentation, the remaining ingredients are added and mixed to the final dough. Dough rest time is generally one hour, with a proof time of 55–63 minutes after dividing and molding. Baking time for a 1-kg loaf is 35–45 minutes.

The production of mixed wheat-rye breads in Hungary is relatively low. Two types of wheat/rye bread are produced in commercial bakeries. The first is based on 70 parts wheat flour and 30 parts rye flour. The second is based on 50 parts wheat flour and 50 parts rye flour. In both cases, a sponge-and-dough procedure is used, with long fermentation time and very low yeast levels (0.25%). The breads are generally hearth-baked.

Another popular Hungarian product is a wheat bread containing potato (as mashed potato or potato flakes). In some cases, mixed wheat-rye bread is used. A typical formula is:

70.5 kg wheat flour (type 1120)
14.1 kg mashed potato
0.3 kg yeast
2.5 kg salt
The potato is added at the dough mixing step.

Vazsonyi bread is a type of Hungarian specialty bread. It was originally made only in certain villages by peasants but now produced commercially in small bakeries. The main raw materials of *Vazsonyi* are wheat flour (type 800), rye flour, and mashed potato (or potato flakes). A typical formula is given below:

197 kg wheat flour (type 800)
22 kg rye flour
6 kg potato flakes
3 kg yeast
5 kg salt
120–140 kg water

A sponge-and-dough procedure is used to produce the bread. For the sponge preparation, 22 parts rye flour and 60 parts wheat flour are used with 90 parts water and 1 part yeast. After relatively long sponge fermentation (5–7 hours), the remaining water, salt, yeast and potato flakes are added and mixed, then the remaining wheat flour is added to create the dough. After dividing and proofing, the bread is hearth-baked. A very tasty end product is obtained which possesses a relatively long shelf life and slow rate of crumb firming. *Soroksari* bread is also a specialty mixed flour bread (70 parts rye flour and 30 parts wheat flour type 1120). It bears strong resemblance to rye bread.

The economic changes brought about by the transition from a centrally planned economy to a market economy have resulted in an increase in the number of small, private bakeries and the assortment of baked goods available to consumers. At present, there are a variety of specialty breads present at the market (e.g., enriched with soya, maize flour, cheese, lentils, linseeds, fiber). "French," "German," and "Italian" bread bakeries and shops have opened.

Bread-type rolls and buns are produced in wide varieties in Hungary. Practically all of them are wheat flour-based products. Three typical Hungarian products in this category are: *zsemlye, kifli* and *fonott kalacs. Zsemlye (Semmel* in German)

Figure 5-2. The process of forming of *fonott kalács,* a Hungarian specialty.

is a small (50g), round lean formula roll based on wheat flour (type 550). Ingredients of traditional Hungarian *zsemlye* are only wheat flour, water (60%), yeast (3%), salt (1.5%). The traditional *zsemlye* is produced by sponge-and-dough procedure. Half of the flour is used for sponge production along with most of the yeast (2.5%). The fermentation time is much shorter (2 hours) than in the case of wheat bread production, but proofing is somewhat longer. The product has a high specific volume, thin yellow crust and elastic crumb. Straight dough procedure is more common in commercial bakeries, along with the use of higher amounts of yeast (3–5%). Addition of emulsifiers is also practiced in commercial production.

Kifli is a fine baked good very popular in Hungary that has a crescent-like shape. Wheat flour (type 550), milk, water, yeast, sugar (2%), fat (1%) are the ingredients. Commercial bakeries use a straight dough procedure. A specific variety of this product is the red pepper crescent.

Fonott kalacs, a knot-like fine baked good, is produced in a variety of sizes ranging from 40 to 500 grams. Its formula is the same as that used for crescent production. The process of

Figure 5-3. Hungarian wheat breads.

forming the final product from fermented dough is shown in Figure 5-2. A selection of Hungarian baked products is shown in Figures 5-3 and 5-4.

Tepertos pogacsa is a product prepared using cracklings (pork). A typical formula is the following one (for 1,000 pieces):

20 kg wheat flour (type 550)
1.4 kg yeast
1.2 kg salt
4.35 kg fat
7 kg cracklings (pork)
25 eggs
0.15 kg pepper

A straight dough procedure is used to produce the dough. After fermentation, the dough is sheeted, rolled, and folded again to develop a puffed pastry multilayered structure.

In recent years, rapid growth has taken place in the production of quick frozen dough pieces for home baking or in-store baking. Pasta products are also well known and consumed widely in Hungary. Hard winter wheat farina is primarily used as raw material. However, an increase in the use of durum wheat for the production of pasta is expected.

Figure 5-4. Hungarian fine breads and rolls.

Table 5-4. Vital Gluten Production of Some Countries in Eastern Europe (1987/1988 Average, 1,000 tons)

Country	Production	Export
Hungary	5.00	2.20
Czechoslovakia	1.50	0.12
Romania	1.05	0.23
Yugoslavia	0.50	...

One special Hungarian pasta product is named *tarhonya*. It is generally home-produced, particularly in villages. White wheat flour and eggs (4 eggs/kg flour) are the raw materials. A dough of hard consistency is prepared and then crumbled to small (1–3 mm) pieces. The crumbled product is air dried and stored until cooking.

Hungary also produces a relatively large quantity of wheat starch and vital gluten. Data on vital gluten production and trade are summarized in Table 5-4.

Poland. With a total production of 25 million tons/year, Poland is one of the largest grain producers in Europe. Rye is the main crop, followed by wheat, barley, oats and triticale. Due to the higher yield of wheat and continuing increase in wheat consumption as part of the population's diet (i.e., the increase in consumption of wheat and mixed rye-wheat breads as opposed to rye bread), it appears that wheat consumption will soon surpass that of rye. In recent years, wheat production has ranged from 6 to 9 million tons (Anon. 1992).

The wheat cultivars grown in Poland may be divided into three groups (Lempka 1970): (a) Western European–type soft winter wheats; (b) red and white winter wheats (classified in Poland as "middle-European" types) of variable baking strength; and (c) hard red winter wheats of the Russian prairie type with good baking quality.

In spite of good baking strength, spring wheat cultivation is not common in Poland, due to lower yield and sensitivity to environmental conditions. Durum wheat is not grown in Poland except for some experimental cultivars such as Pulawska and Twarda.

The laboratory baking test is the key factor in determining the baking quality of wheat flour in Poland. Determination of wet gluten content and its quality, as well as α-amylase activity (Falling number) are also common measures.

A wide variety of baked goods made from soft wheat flour (including cookies, cakes, wafers, biscuits, crackers, sweet goods, etc.) are produced in Poland. *Herbatniki* is a chemically leavened cookie (biscuit)-like product, produced from soft wheat flour, sugar, fat, milk, baking powder, salt and flavorings. Three types are known: sweet (soft), half sweet, and hard (produced without sugar). Length and width of the products are up to 80 mm. and thickness around 6 mm.

Pierniki is produced from soft wheat flour (occasionally from wheat-rye flour mixture) plus sugar or honey, malt extract, raisins, flavorings but without fat, milk or eggs. It is sometimes filled or coated with chocolate. *Biszkopty* and *makaroniki* are foam-type cakes. The main ingredients are wheat flour, sugar and egg white. *Makaroniki* always contains roasted oilseeds.

Breads made from a blend of hard wheat and rye flours are the main products of bakeries in Poland. Six types of mixed flour breads are produced (Lempka 1970, Ambroziak 1988, Pomeranz 1987), differing in rye-wheat flour ratio from 70:30 to 20:80. *Sandomierski* is a sour-type rye-wheat bread. It is produced using a mix of 30 parts wheat flour (type 800) and 70 parts rye flour. The fermentation is a long multistage sour. It is almost always hearth-baked in round or oblong shapes and weighs between 1 and 2 kg. It has a relatively high acidity and resembles typical rye bread.

A popular mixed flour bread is the *praski* bread, which is baked from 60 parts wheat flour (type 800) and 40 parts rye flour. A two- or three-stage sour fermentation is used along with baker's yeast. The dough is baked in 1.0- or 1.5-kg oblong loaves or in pans. The final product has a milder sour taste and a relatively higher volume than *Sandomierski*.

The mixed flour breads *naleczkowski* and *leczycki* are produced in smaller quantities, in limited regions of the country. They are baked from 50 parts wheat flour (type 800) and 50 parts rye flour. Both types of breads are yeast leavened, produced by a two- or three-step sponge method. The typical formula is as follows:

    Sponge: 48 parts rye flour
             45 parts water
             0.5–1.0 parts yeast
             fermentation time 5.5–6.5 hours (28–30°C)
    Dough: 50 parts rye flour
             50 parts wheat flour
             10–12 parts water
             1.5 parts salt
             fermentation time 0.5–1.0 hour (28–30°C)

Sponge fermentation is about 5.5–6.5 hr at 28–30°C. After mixing other ingredients, dough is fermented for 30–60 minutes, followed by dividing, molding and a short proofing. The final products are mainly hearth-baked and have the characteristics of a wheat bread.

Zakopanski is a traditional specialty bread from the mountainous region in South Poland. The bread is named after the well-known skiing resort Zakopane. *Zakopanski* is made from equal parts of rye flour and wheat flour (type 550) by the sponge-and-dough method. A rye sponge is prepared using about 40 parts of skimmed sour milk per 100 parts of rye flour. The remaining ingredients are added at the dough stage. Round shaped 0.5 kg loaves are formed and fermented on wooden boards. The final product has a thick, hard crust and a characteristic taste.

Mazowiecki is a mixed flour bread containing a high proportion of wheat flour (80 parts, type 800) and only 20 parts of rye flour. This type of bread is unique because of the variety of forms in which it is produced. The best *mazowiecki* is produced from a yeast-leavened wheat flour dough (Pomeranz 1987). Wheat flour (40 kg) plus 33 kg water and 1.5–2.0 kg baker's yeast is used to prepare a sponge which ferments for 3–3.5 hr at 27°C. The rye sour is produced from 18 kg rye flour and 13 kg water (29–30°C) in which the salt (1.5–1.7 kg) is dissolved. This is added to the sponge, along with 40 kg of wheat flour, and the whole is mixed thoroughly and fermented another 30 minutes. After scaling and forming, the rounded pieces are proofed for 35–45 minutes, during which time loaves are lightly moistened twice, once immediately after forming and the second time just before baking. The loaves are scored slightly before the second moistening and dusted with poppy seeds. Baking is in ovens saturated with steam. The loaves are moved 15 minutes after placing in the oven. The initial baking temperature should not exceed 220–230°C. Total baking time is 38–45 minutes for a dough piece weighing about 935 g. *Mazowiecki* is baked as oblong hearth loaves or in pans.

A number of baked goods are produced from 100% wheat flour. The most typical types are hearth-baked or pan breads, white breads, and a variety of different rolls. Both sponge-and-dough and one-step straight dough procedures are used in production. A typical formula for common sponge/dough white bread is the following:

Sponge:	45 parts wheat flour (type 500)
	45 parts water
	1.0–1.5 parts yeast
Temperature:	24–27°C
Fermentation time:	3.0–3.5 hours
Dough:	43 parts wheat flour (type 500)
	8–10 parts water
	1.0–1.5 parts salt
	1.0 part sugar
Temperature:	28–32°C
Proof time:	20 minutes

In the case of rolls, the most typical formula is:

Sponge:	53 parts wheat flour (type 500)
	37 parts water
	1.5–2.0 parts yeast
Temperature:	26–27°C
Fermentation time:	3.0–3.5 hours
Dough:	45 parts wheat flour (type 500)
	12 parts water
	2.0 parts sugar
	3.0 parts shortening
	1.3–1.7 parts salt

Lecitinowy is a special wheat bread containing lecithin, dried skim milk, and shortening. A typical formula for production of *Lecitinowy* is as follows:

99 kg wheat flour (type 500)
1 kg rye flour
4 kg dried skim milk
1 kg lecithin
5 kg sugar
5 kg margarine
1.5–2.0 kg yeast
0.8–1.0 kg salt

A two-step sponge-and-dough procedure is used in production. Sponge is made from 50 parts wheat flour, 40 parts water and 2 parts of yeast then fermented at 28°C for 3.5 hours. All other ingredients are dissolved in water and, together with the remaining wheat flour, added to the sponge and mixed to final dough. After dividing, molding and proofing for 40 minutes, it is baked, generally in pans.

French-type baguettes and graham bread are also produced in Poland. Among the latter type of specialty breads, *blonwit* bread has the following formulas:

60 parts wheat flour
40 parts whole wheat
10 parts wheat bran
2 parts yeast
4 parts dried skim milk
5 parts wheat germ
2 parts dried food yeast
1.5–1.7 parts salt

A sponge-and-dough procedure is *blonwit* bread production. The bread is baked in pans.

Pasta products are also made in Poland. Durum wheat flour or hard wheat flour (type 500) is used as raw material. Three types of products are known: (1) those without eggs, (2) those containing eggs, and (3) enriched pasta products.

Romania is among the important cereal and wheat producers in Eastern Europe (Table 5-5), with an annual wheat production averaging near 10 million tons. Maize is the main cereal grown in Romania with a production of 10–14 million tons per year. Production of rye and oats is very small. Wheat is generally used for food purposes, and the surplus is also exported.

Beginning in the 1960s, Romania, as well as other Eastern European countries, made concerted efforts to select newer cultivars with higher yield. These were based on Soviet, Italian, and Bulgarian varieties. As a result, wheat production increased, but the newer cultivars are generally less cold-resistant and have widely varying baking qualities.

Table 5-5. Cereal and Wheat Production in Romania (1,000 tons)

Year	Cereals	Wheat
1979–1981	19,827	5,471
1979	19,337	4,676
1980	20,200	6,427
1981	19,945	5,310
1982	22,334	6,460
1983	19,606	5,205
1984	23,577	7,580
1985	23,049	5,666
1986	30,372	7,320
1987	31,604	9,672

Source: FAO Statistical Yearbook, 1990

Historically, maize has played the largest role in the nutrition of the Romanian population, and mixed wheat-maize bread is widely consumed. There is, however, a growing tendency toward the consumption of white wheat breads and rolls. Production and consumption of wheat-rye breads is limited mainly to some areas of Transylvania.

Commercial production of what might be termed typical soft wheat products and pasta products is relatively low in comparison with Western Europe.

In addition to use in flour-based foods, some wheat is used for the production of wheat starch and vital gluten production.

Yugoslavia. As a result of large political and social changes in Eastern Europe, and particularly in the Balkan Peninsula, multiple new states were formed from the territory of what was Yugoslavia (originally a federation of six republics). Due to difficulties in the receipt of specific information about the new states, it is nearly impossible to give a separate overview for each new state. As a consequence, the majority of data given in this segment relates to what was the former Yugoslavia. Nevertheless, where possible, data relating to new countries (Slovenia, Serbia, Croatia, Bosnia-Hercegovina, and Macedonia) are also included.

Although wheat is grown in all the parts of former Yugoslavia, the main cereal-producing area is the northern part of the country (the Danube basin and Slavonia). Some statistical data concerning wheat growing area, production and yield is summarized in Table 5-6 (SZS, 1991). Clearly, corn and wheat are the main cereals. The rye production is low and continues to decline.

Currently, there is no uniform procedure for wheat classification in the different parts of former Yugoslavia. During the late 1980s, a proposal was made for such a classification, and the following classes were proposed (Saric et al., 1987): bread wheat with three subcategories (improved high-quality wheat, normal bread wheat, low-quality bread wheat), soft (biscuit) wheat and feed wheat.

Table 5-6. Wheat Production in Yugoslavia

Year	Area (1,000 ha)	Production (1,000 tons)	Yield (kg/ha)
1979	1,684	4,512	2,960
1989	1,479	5,599	3,780
1990	1,498	6,350	4,250

Source: Statistical Yearbook of Yugoslavia, 1991

Most of the wheat grown in the different parts of Yugoslavia is winter wheat. About 15–20% of total production is a hard, high-quality type (e.g., the cultivar Partizanka). Most of the cultivars grown in Yugoslavia are intermediate, between hard and soft wheats, with variable baking quality (e.g., the cultivars Novosadska, Rana, and Balkan). Among the soft wheat cultivars, Super Zlatna and Baranjka are the best known. Durum wheat production is not important in Yugoslavia.

Determination of baking quality in the region relies on sedimentation value, protein content, flour yield, and baking test results. In the case of soft wheat quality, the sedimentation value, alkaline water retention capacity, and flour suspension viscosity are used. To lesser extent, farinographic and extensographic parameters are used as indicators of quality.

White (refined) wheat flour production is approximately 3.2–3.4 million tons/year. Of that total, 90% is supplied to the baking industry and the remainder to pasta and confectionery production. Roughly half of the flour is used for bread production, 45% for specialty bread and 5% for rolls and other products.

In all of Yugoslavia, wheat bread is the main food product made from cereals. Rye and maize are used only in smaller amounts, and then only in some republics (rye in Slovenia, maize in Serbia). The typical wheat bread is made from white (refined) wheat flour. Both sponge-and-dough and straight dough methods are employed to produce bread. In the villages of Croatia and the Adriatic coast, a short sponge is used. In the central Yugoslavia, a much longer "overnight" sponge is common. Villages in Serbia, Macedonia and Croatia use an air-dried sourdough piece as a starter in bread production. Among the 40 types of breads produced in the region, the oblong, hearth-baked *vekna* and the round *pogatcha* are typical. Ascorbic acid is the only flour improver permitted in all republics. On the other hand, the use of emulsifiers is very common in commercial white wheat breads and rolls production.

In some areas, an extract from *Carum carvi* is added to bread to improve taste and flavor. In Macedonia, some wheat breads contain an extract of hops (*Humulus lupus*) as flavoring.

Mixed wheat flour-rye flavor breads are produced in Slovenia. Wheat-maize bread is produced in the villages of Serbia and Bosnia-Hercegovina. *Mlinci* is a thin chapatti-like bread of mixed flour produced in Croatia.

Breakfast rolls and fine baked goods (about 100 varieties in all) are widely produced and used. Among the specialty products are the Serbian *kolatsch*, a sweet baked good prepared from refined wheat flour, milk, eggs, sunflower oil, sugar, salt and yeast baked in a round shaped pan. The surface of the *kolatsch* is moistened with egg yolk and decorated with ornaments made of dough. This product is prepared for various religious feasts, and the ornaments correspond with a specific feast.

Gibanitza is prepared from a refined white wheat flour-based dough. Thin sheets are formed, covered with oil or fat and layered on top of each other with curd and/or cheese used as a filler. This is a salted product as opposed to *tchesnitza*, which is a sweet product containing layers of honey and ground nuts. In Serbia, *tchesnitza* is baked for Christmas and New Year. A similar product in Serbia and Bosnia-Hercegovina is known as *burek*.

Soft wheat products (of all types) are known throughout Yugoslavian regions, but the main producing region is Croatia. Cookies and biscuits are the most popular of the products.

Pasta is produced in the country. The most common starting material is hard winter wheat farina, but some durum wheat semolina is also used. Most of the types and shapes known internationally are commercially produced.

References

Ambroziak, Z., 1988. Technologia piekarstwa (Bakery technology). Wydawnictwo Szkolne i Pedagogiczne, Warszawa.

Anon., 1987. Focus on Hungary, World Grain 5:29.

Anon., 1990. Focus on Yugoslavia. World Grain 8:29.

Anon., 1992. Focus on Poland, World Grain 10:24.

Dodok, L., 1988. Chemia a technologija trvanliveho peciva. Vydavatelstvo Technickej a Ekonomickej Literatury., Bratislava.

Gasztonyi, K. and Schneller, M., 1963. Sutöipar (Bakery Industry), Muszaki Könyvkiadó, Budapest.

Központi Statistikai Hivatal (KSH), 1983–1992. Magyar Statisztikai Zsebkönyv (Hungarian Statistical Pocket Book), Statisztikai Kiadó Vállalat.

Lekes, J., 1983. Intensification of Cereal Production in Czechoslovakia and Increasing of its Nutritive and Processing Quality. In: Developments in Food Science Vol. 5.A., J. Holas and J. Kratochvil (Eds.), Elsevier Publ., Amsterdam, p. 313.

Lempka, A., 1970. Towaroznanstwo produktow spozywczych, P.W.E. Publ. Warszawa.

Pomeranz, Y., 1987. Modern Cereal Science and Technology, VCH Publ., Weinheim.

Popov, P., Simeonov, B. and Matchev, S., 1969. Problems of Breeding and Agrotechnics of Soft Winter Wheats (in Bulgarian), Acad. Sci. Publ., Sofia.

Sabo, B. and Kozel, L., 1983. The role of Cereals in Contemporary Nutrition. In: Developments in Food Science, Vol. 5B, Holas J. and Kratochvil, J. (Eds.) Elsevier Publ., Amsterdam, p. 1133.

Saric, M., Gavrilovic, M., Sekulic, R., Jovanovic, O. and Dozet, J., 1987, Sorta u borbi za kvalitetnu sirovinu, Zito-Hleb 14:23.

Savezni Zavod za Statistiku (SZS), 1991, Statisticki Kalendar Jugoslavije, Beograd, p. 66.

Statisticka Rocenka Ceske A Slovenske Federativni Republiky (Statistical Yearbook of Czech and Slovak Federal Republic), 1990, SNTL Publ., Praha.

CHAPTER 6

Wheat Usage in Northern Europe

Hannu O. Salovaara and Kjell M. Fjell

Introduction

Among the cereal grains cultivated in the Nordic countries (i.e., Denmark, Finland, Norway and Sweden), wheat ranks second to barley in total harvest. High yield and versatile end uses have contributed to a substantial increase in the wheat crop in Northern Europe, especially in Denmark in the last two or three decades. At the same time the significance of oats and rye, also very typical of the area, has diminished. The climatic conditions vary within Scandinavia and determine the type of wheat grown. Only common wheat of the European type is produced. No durum wheat is grown in the area.

Today, less than one quarter of the total Nordic wheat harvest is used for human consumption. The main use of wheat is as animal feed, either directly on the farm or as a component of industrially produced animal fodder. However, the local usage of wheat in the four countries is dissimilar and greatly dependent not only on the climatic conditions for local production but also on differences in local agricultural policies. The latter may be subject to fundamental changes in the 1990s in Sweden, Finland and Norway due to the progress of European economic integration.

In the past, Denmark and Sweden have exported wheat regularly and Finland occasionally, whereas the volume of the Norwegian crop has never met the domestic need. Apart from Norway, most of the wheat used in the Nordic countries comes from local production. However, there are certain end uses for wheat, such as pasta and biscuit production, which may require wheat qualities not met by the local crop. Further reasons for wheat import to the Nordic countries are occasional sprouting problems and other defects in the quality of the local wheat harvest.

Most of the wheat used for food in the Nordic countries is milled into bread flour for the baking industry, small bakeries, and home bakers. Additional end users of wheat flour, which may require wheat flours of specific qualities, include biscuit and macaroni manufacturers, and the confectionary industry. Denmark and Sweden have wheats well suited for biscuit and confectionary purposes, but the quality of the blends may be improved by adding imported soft wheats.

Harvest and Yield of Wheat

Numeric data on recent harvests and yields of spring and winter wheat in the Nordic countries are shown in Table 6-1. The harvest of wheat in the Nordic countries currently totals more than 5.5 million tons yearly. The Danish wheat crop alone, more than 3.0 million tons, exceeds that of all the other Nordic countries combined. Nearly one half of the Nordic wheat growing area is in Denmark (more than 400,000 hectares). The total wheat area and harvest of the Nordic countries account for approximately 6 and 8% of the bread wheat area and harvest in the European Community (EC), respectively.

In the last twenty years, wheat production has more than doubled in the Nordic countries, due in particular to the expansion in Danish wheat production, which has increased fourfold since 1970, and has benefited from Denmark's membership of the EC. Practically all the wheat grown in Denmark is winter sown, as is most Swedish wheat, whereas half of the Norwegian wheat crop and most of the Finnish crop are of the spring-sown type. The average yield per hectare of winter wheat in Denmark equals the highest European averages of 7–8 tons, followed by 5–6 tons in Sweden, 4–5 tons in Norway and 3–4 in Finland. Spring wheat yields 0.5–2 tons less than winter wheats. In Nordic climatic conditions, cultivation of winter wheat is restricted by problems of winter hardiness. Finland, Sweden and Norway are practically the only countries in Europe where spring wheat is grown.

The prices paid to the farmers for wheat in Finland and

Table 6-1. Total harvest and yield per hectare of winter and spring wheats in Northern Europe

	Denmark		Finland		Norway		Sweden	
	1988–1990 average	1991	1988–1990 average	1991	1988–1991 average	1991	1986–1990 average	1991
Harvest (1,000 tons)								
Winter wheat	3,024	3,895	81	149	25	103	1,399	1,325
Spring wheat	62	58	392	281	145	143	316	156
Total	3,086	3,953	473	430	170	246	1,715	1,481
Yield (kg/ha)								
Winter wheat	7,180	7,430	3,720	3,670	4,560	5,150	5,880	5,890
Spring wheat	5,030	5,270	3,130	3,620	3,860	4,350	4,330	4,730

Sources: Denmark: Eurostat 1992; Finland: Finnish Board of Agriculture; Norway: Statkorn (Norwegian Grain); Sweden: JEM, Jordbruksekonomiska meddelanden 1992.

Table 6-2. Wheat export and import from and to the Nordic countries (1,000 metric tons)

	Denmark		Finland		Norway		Sweden	
	1987–1990 average	1991	1988–1990 average	1991	1990	1991	1988–1990 average	1991
Export	605	1,105	25[a]	27[a]	0	0	472	740
Import, total	116	52	87	26	159	130	56[b]	46
Common bread wheat	d[c]	d	...	...	71	48	...	...
US/Canadian wheat	d	d	69	(26)	77	63	41[b]	31
SW, soft	d	d	7	12	...	5	...	...
Durum	d	d	11	14	11	14	15[b]	15

[a] Export under international food aid program.
[b] Approximate values.
[c] Detailed data not available.

Sources: Denmark: Eurostat; Finland: Finnish Grain Board; Norway: Statkorn (Norwegian Grain); Sweden: JEM, Jordbruksekonomiska meddelanden 1992

Norway are currently much higher than the price paid in the EC or the price prevailing in the world market. The local price results from the local policy, which has had the aim of ensuring a certain domestic production and self-sufficiency. The high price is meant to compensate farmers for lower yields resulting from the climatic conditions.

Export and Import

As a consequence of their increased production, Denmark and Sweden have become regular exporters of wheat, whereas Norway has always been a net importer, as has Finland in the long term. Denmark exports more than 1.0 million tons yearly (Table 6-2). Exports from Denmark go both to the internal EC wheat market and to the world market. The Swedish surplus is sold in the world market and finds its way to a number of countries, including Norway. In addition to the wheat export shown in Table 6-2 Sweden has, under the World Food Program, exported 50,000–60,000 tons of wheat flour (79–80% extraction rate) yearly made from domestic Swedish winter wheat. This activity, supported by the Swedish government, has constituted up to 10–12% of the total production of Swedish flour mills.

Import and export of wheat are a state monopoly in Norway and Finland, and under state control in Sweden, too. Table 6-2 shows data on wheat imports by the Nordic countries. The majority of the imported wheat has been of North American origin. Denmark also imports some EC wheat, as does Norway, along with Swedish, Argentinian or even Saudi Arabian wheat. The imported wheats are used mainly for breadmaking, as strong or medium-strong components in the wheat blends. In the Norwegian blends weak (soft) wheat mainly from Europe is also used to adjust breadmaking quality. Other uses of imported wheat are biscuit manufacture and macaroni or pasta production.

Quality of Nordic Wheat

The average quality of wheat harvested in the Nordic countries is shown in Table 6-3. The quality of Nordic wheat is subject to local and annual variations, so average figures may sometimes be misleading. The protein content of Nordic winter wheat seldom exceeds 12–13% (dry basis), whereas spring

Table 6-3. Some quality characteristics of winter and spring wheat grown in Northern European countries

	Denmark		Finland		Norway		Sweden	
	1988-1990[a]	1991	1988-1990[a]	1991	1988-1990[a]	1991	1988-1990[a]	1991
Protein content, %[b]								
Winter wheat	11.6	d	12.2	11.3	11.6	11.7	11.4	11.2
Spring wheat	d[c]	d	14.0	13.2	13.9	13.9	13.2	13.1
Falling number, s								
Winter wheat	307[d]	389[e]	208	170	299	259	>190[f]	180
Spring wheat	d	d	267	271	280	281	>190[f]	200
Hectoliter weight, kg								
Winter wheat	d	d	80.2	80.1	81.8	81.1	81.7	82.9
Spring wheat	d	d	80.7	81.1	81.2	81.4	80.7	80.7

[a] Average.
[b] Protein 5.7 × N, dry matter basis.
[c] Detailed data not available.
[d] 1988-1989 average.
[e] Actually average of samples representing three important bread wheat varieties.
[f] More than 90% of the analyzed samples exceeded falling number 190.

Sources: Denmark: The Association of Danish Millers; Finland: Finnish Grain Board; Norway: Statkorn (Norwegian Grain); Sweden: JEM, Jordbruksekonomiska meddelanden.

wheat may have an average protein content of 13–15%. Farmers are encouraged to produce high-protein wheat by using a two-step nitrogen application. A protein pricing system is applied in Sweden, Finland and Norway. The premium for higher protein to the farmer also makes spring wheat growing somewhat more competitive in comparison to winter wheat. In domestic wheat trade in Sweden and Norway, spring wheat goes under the term "quality wheat," which relates to its higher protein content and other properties.

Practically all wheat cultivars in the Nordic countries are hard kernel types, and hence of good milling properties (Svensson 1987). Vitreousness is not a property present in Nordic wheats. Winter wheat varieties with soft kernel or non-baking dough properties are not commonly cultivated, although sometimes they show up as examples of European high-yielding wheats.

The baking properties of Nordic wheat varieties originate in part from efforts in the first half of the century and the early breeding lines used by local wheat breeders (Svensson 1987, Juuti 1988). Modern electrophoretic analyses have shown that the wheat varieties grown in Northern Europe possess HMW-glutenin subunits characterized by high quality scores of 8–10 in Payne's classification (Sontag et al. 1986, Svensson 1987, Mjaerum 1992, Uhlen and Mosleth 1992). On the other hand, Svensson (1992) suggested that the optimum glutenin score for good quality with respect to Swedish breadmaking practice may not be higher than 7–8.

The most severe quality problem in Nordic wheat production is the risk of preharvest sprouting (excessive alpha-amylase activity) in years when rainy periods occur at harvest time in August–September. Spring wheat has this problem more frequently because of its later harvest. There have been years when a considerable proportion of the Nordic spring wheat crop has been rejected by the flour milling industry because of sprout damage. Therefore, an important goal of Nordic plant breeders has been to increase the sprouting tolerance (dormancy) of the wheats. Another solution would be the breeding of varieties with shorter growing time and, hence, earlier harvest. However, shorter growing time is generally related to lower yield. Currently the time in days from planting to harvest of Nordic spring wheats in the Nordic climatic conditions varies from 100 to 108 days (Juuti 1988).

Because of the prevalence of sprout damage the falling number has been an important tool in selecting wheat suitable for flour milling ever since this method became available in the 1960s after the long developing process carried out by Hagberg of Sweden (Perten 1992). Without such a practical tool it might not be possible to cultivate wheat in Northern Europe to the present extent. The falling number is used extensively both as a quality parameter in pricing farmers' grain, and in segregation of wheat for flour milling.

Consumption of Wheat

The per capita consumption figures for wheat as food in Denmark, Finland and Sweden are among the lowest recorded in Europe and the OECD countries. Table 6-4 shows wheat consumption figures in comparison with other domestic cereal grains. Typical of Norway is higher wheat consumption and lower rye consumption as compared with the neighboring countries. However, the relatively low wheat consumption in Denmark, Finland and Sweden is compensated in part by the relatively high consumption of rye, which is used not only for making crisp bread and whole rye bread but also for variety type "mixed" or brown breads containing some rye flour.

The declining trend in the total consumption of cereal grains in the Scandinavian countries has been the subject of concern not only in the cereal industry but also among nutrition experts. In Sweden campaigns directed at consumers were launched to increase consumption of cereal grains substantially. The programs have been financed by local flour milling and baking industries. Similar although less extensive activities exist in all Nordic countries. The campaigns in Sweden reversed that country's declining consumption trend. In Norway,

Table 6-4. Human consumption of wheat in comparison with rye, barley, and oats in Nordic countries (kg/yr per capita)

Commodity	Denmark	Finland	Norway	Sweden
Wheat				
Grain basis	56.0	58.4	80.8	68.6
Flour basis	43.1	43.8	64.6	48.0
Rye				
Grain basis	20.2	19.8	9.3	15.1
Flour basis	17.2	19.4	7.4	12.5
Barley				
Grain basis	0.1	2.0	0.4	0.4
Flour basis	0.1	1.5	0.2	0.3
Oats				
Grain basis	4.8	5.3	2.8	3.6
Flour basis	2.5	3.2	1.4	2.2

Source: OECD Food Balance Sheets, Paris 1991

a government program was launched in 1974–1975, forecasting 90 kg per capita consumption of flour by 1990. This goal was not reached, as can be seen from Table 6-4. One reason for this is the removal of heavy subsidies on flour at the beginning of the 1980s. The subsequent price increase eliminated the flour "leakage" across the Swedish border, decreased the waste and the use of flour for feed.

End Uses of Wheat

Table 6-5 lists the volumes of the principal industrial end uses for wheat in the Nordic countries. Wheat used for food amounts to approximately 1.4 million tons yearly, less than five percent of that in the EC (Eurostat 1992). In Denmark more than 80% of the domestic wheat usage goes to animal feed either directly on farms or via the animal feed industry. In Sweden one third of the domestic wheat usage is for feed (Johansson 1989). In Finland and Norway, where all wheat is grown for the purpose of flour milling and breadmaking, usage of wheat for other purposes arises only when the quality of wheat is too low for breadmaking, especially due to sprouting damage. Other important food uses in addition to regular flour milling are biscuit and macaroni manufacturing. Both require specific flours made at least in part from imported wheat. The Swedish wheat starch and gluten industry uses domestic wheat. Wheat of high falling number is required in the pro-

Table 6-5. Industrial end-uses of wheat in Northern European countries (1,000 tonnes; export and seed excluded)

End-use	Denmark 1991	Finland 1991	Norway 1991	Sweden 1991
Wheat for food uses, total	291	275	337	614[a]
Flour for bread and biscuit manufacturing	225	254	300	450
Flour for pasta and macaroni industry	. . .	17	11	15
Wheat for starch and gluten industry	. . .	. . .	. . .	20
Wheat for animal feed	1,619	42	76	350
Other uses	0	32[b]	21[c]	. . .

[a]Includes wheat used for milling 54.000 tons of wheat flour exported under food aid program.
[b]Includes wheat used for fur-animal feed and alcohol production and 2.700 tonnes for candy licorice production.
[c]Includes some flour used for flat bread production; other uses under this figure: breading flour for fish fingers, etc., and technical purposes.

Sources: Denmark: Eurostat and The Association of Danish Millers; Finland: Finnish Grain Board; Norway: Statkorn (Norwegian Grain); Sweden: JEM, Jordbruksekonomiska meddelanden

duction of fish fodder and feed for fur animals. Wheat for this purpose and other feed purposes has been exported to Norway and to Finland.

Quality Criteria for Wheat Flour. The production of flour of good and consistent breadmaking properties requires wheat of appropriate falling number, protein content, hectoliter weight, and other properties. Consequently the grist for milling bread flour is often composed of two or more different quality classes. The precise combination depends on the quality of wheat classes available.

Examples of some typical quality requirements for domestic and imported wheats applied by Nordic flour mills are shown in Table 6-6. The quality of individual lots of wheat purchased from the domestic market can vary over a wide range. The capability of the flour mills to use a wheat lot of lower quality depends on their access to compensating high-quality wheat lots of domestic or imported origin. Because of the decisive role of governmental grain boards in wheat storage, export and import in Norway and Finland, a close cooperation exists between these organizations and commercial flour mills in order to maintain a balance and continuity of wheat and flour quality.

Flour millers may also have special requirements depending on the quality of wheat available, and the demands from their customers (Fjell 1992). Therefore, the quality criteria shown in Table 6-6 may vary in many ways. In general, it appears that the quality criteria applied in Norway are higher than elsewhere. This may be because Norwegian flour mills are accustomed to a high percentage of imported strong wheat supplied by the state-owned grain board. On the other hand the flour mills (and bakeries) in Finland seem to have adapted to somewhat lower quality requirements, especially to a lower falling number.

Most problematic with respect to end use value are wheat lots of low falling number. Flour mills in Finland have sometimes used wheat lots of falling number below 160, whereas the rest of the Nordic mills use a falling number of 190 as their lowest acceptable limit for domestic wheat. In Sweden, Finland and Norway the price of wheat paid to the farmer goes down progressively with decreasing falling number. There have been years when a considerable proportion of the wheat harvest in these countries has been unsuitable for flour milling due to sprouting damage as measured by its low falling number.

Quality requirements for the soft white wheat imported for

Table 6-6. Some quality criteria applied by Nordic flour mills for wheat for specific end-uses

	Breadmaking wheat			Biscuit	Macaroni
Quality criteria	Domestic wheat (lowest acceptable)	Imported wheat	Specification for the grist to be milled[a]	Imported soft wheat[b]	Imported durum[c]
Protein content, % d.m.	>12.0 (10.5)*	>11.0–16.0	>13.0	10–12	>15.0
Hectoliter weight, kg	>76	>80	>78	>80	>80
Falling number, s	>190 (>160)	>300	>250	>300	>300
Gluten content, wet w.	...	>25	>27	20–25	...
Other grain, %	<1	<1	...	...	...
Other characteristics	...[d]	...	...	...	...

*Figures in parentheses are for winter wheat.
[a]These criteria must be met by the wheat grist to be milled into breadmaking flour; some individual wheat lots approved by the mill may deviate from these aimed values.
[b]Soft White wheat. Gluten must be soft. Dough development time <1 min. Degree of softening >130 in 120 min. Absorption <55.0%.
[c]Durum wheat, preferably North American. Amber in color. Soft kernels <8%.
[d]Green, immature kernels not more than 5% (Finland).

biscuit manufacturing and for durum wheat include hectoliter weight above 80 kg and falling number above 300. Protein content of soft wheat for biscuits should not exceed 12% (dry basis), whereas that of durum wheat should exceed 15%. In most cases North American Soft White and Durum wheats are preferred by the industry.

Typical values for some quality parameters for regular breadmaking flours, as generally applied in the Nordic flour mills, are shown in Table 6-7. The values shown originate from information received from individual mills and local milling associations. Official standards for flour quality exist only in Norway, where the commercial mills process and market the many types of flour on behalf of the governmental grain board. Unofficial agreements on protein, dough yield, etc. are not uncommon between the Nordic mills and their customers. Baking industry experience has shown that flours of good industrial baking performance in French bread processes had high wet gluten content, high farinograph absorption and farinograph characteristics generally interpreted to indicate high baking quality (Salovaara 1987). Special flours with low ash content (below 0.55%, dry basis), coarse particle size, and no ascorbic acid are made for confectionary use. Flour with no ascorbic

Table 6-7. Typical quality characteristics for regular wheat flour applied in Nordic mills

| Specification | White flour | | Brown flour |
	Typical levels	Norwegian criteria	
Moisture, %	14.5–15.0	≤15.0	14.5–15.0
Ash, %	0.65–0.75	0.6 ± 0.05	1.1–1.3
Wet gluten	27–30	...	...
Protein, % (d.m.)	11.5–14.0[a]	13.4 ± 0.3	...
Falling number, s	240–300	>250	200–270
Ascorbic acid added, g/100 kg	1.5–4	3–4	5–8
Farinograph characteristics			
Absorption, %	56–62	59 ± 2	55–65
Dough development time, min	2–4[b]	...	...
Stability, min	4–8[b]	...	...
Zeleny-value	>30	...	...
Particle size (examples)			
Through (sieve opening μm)[c]	50–60% (95 μm)		>80%
	55–65% (75 μm)		
Test baking			
Specific loaf volume, ml/g[d]	4.0–7.0	6.5 ± 0.3	...

[a]The required protein content may vary depending on the quality of local and imported wheat and on end-use.
[b]Farinograph characteristics may vary according to protein content.
[c]Flour granularity varies locally and with end-use.
[d]Specific loaf volume (ml/g of bread) varies with protein content and other quality parameters.

Table 6-8. Domestic wheat flour production and usage in industry and households in the Nordic countries in 1991

Production and usage	Denmark		Finland		Norway		Sweden	
	1,000 tons	%	1,000 tons	%	1,000 tons	%	1,000 tons	%
Wheat flour production, total	256	100	230	100	269	100	436	100
Bakery and industrial usage	215	84	150	65	188	70	327	75
Home-baking flour	41	16	80	35	81	30	109	25

Sources: Denmark: The Association of Danish Millers; Finland: Finnish Grain Board; Norway: Statkorn (Norwegian Grain); Sweden: JEM, Jordbruksekonomiska meddelanden

acid supplementation is also used in some Danish pastry processes.

Wheat Flour Usage. Most wheat flour is used by bakeries for making various types of white breads, brown breads, sweet coffee breads and confectionary products. Besides these industrial uses, one quarter to one third of all wheat flour produced in Finland, Norway and Sweden is sold for household use (Table 6-8). Different types of flour products and flour mixes for home bakers are on the market and intensively advertised, even on TV. Popular baked goods made at home are white bread, brown bread and brown rolls, yeast raised sweet coffee rolls, and cakes. Most home baking flour is similar to the flour delivered to the commercial baker. In Finland a typical home baking flour for sweet roll and cake baking is characterized by lower ash content (below 0.6% dry basis), higher falling number (over 300) and coarser particle size as compared to bakers' regular white flour.

Home baking is more common in the sparsely populated northern parts of Norway, Sweden and Finland than in the southern parts, where the consumers have more access to bakeries. However, home baking seems to survive even in urban areas. The economy of breadmaking at home together with other factors, such as specific nutritional demands or idealistic motives may inspire certain consumers to home baking and sustain the market for home baking flours. In addition, automatic bread machines have found a good market in the Nordic countries and broadened the spectrum of home baking households.

Production of Wheat-Based Goods

Table 6-9 shows total volumes of bakery goods and other industrial foods made from wheat flour in the Nordic countries. Soft, yeast-leavened bread is the major product. White breads,

Table 6-9. Production of wheat-based or wheat-containing foods in the Nordic countries (baked goods and pasta), 1,000 tonnes, for domestic consumption

Commodity	Denmark	Finland	Norway	Sweden
Baked goods				
Bread, soft (wheat and rye)	206	168	188	263
Crisp bread/flat bread[a]	6	18	5	51
Sweet coffee rolls, etc.	37	54	[b]	31
Cakes, confectionary		8	22	32
Biscuits, zwieback, etc.	63	15		44
Other products:				
Macaroni, pasta, and others	16	29	13	79

[a]Mostly made from rye.
[b]Included in Bread, soft.

Sources: Denmark: 1991 Statistical Bureau; Finland: Valmisteverotilasto 1991, Finnish Food Industry; Norway: 1991 Statkorn (Norwegian Grain); Sweden: JEM, Jordbruks-ekonomiska meddelanden 1992

such as the so-called French bread, white rolls and baguette-type breads constitute approximately 10% of the total bread production. Popular in all the countries are various types of mixed or "variety" breads made from white wheat flour mixed with brown or whole wheat flour, wholemeal rye, oat flakes, broken kernels, etc. In Denmark and Finland approximately 30% of the soft bread production is soft wholemeal rye bread, which may contain some brown wheat flour. Yeast leavened crisp breads are typical of the area, especially in Sweden and Finland. Most of them are made from wholemeal rye. The unyeasted Norwegian flatbreads contain rye and other grains, including wheat. Typical of Northern Europe are the sweet, yeast raised coffee breads, such as sweet rolls, buns, and pastries made from white wheat flour. Their production constitutes 10–20% of the total bakery production in these countries. Most cookies are made from flour milled from imported soft white wheat, whereas crackers are made at least in part from bread wheat. A considerable proportion of the production of several cookie manufacturers goes to the export market, and the industry is dependent on good quality soft wheat imported mainly from the U.S. Both the production and consumption of macaroni and pasta have been increasing in the past ten years, especially in Sweden. Most of the locally made macaroni products are made from imported durum wheat.

Typical Wheat-Based Products

Table 6-10 shows formulas and process parameters for a few typical wheat-based bakery products popular in Northern Europe. Typical of the baking processes are relatively intensive

Table 6-10. Examples of typical formulas and some important process parameters for popular wheat-based bakery products in Northern Europe

	French bread "franska"	Brown wheat bread	Swedish sweet brown bread "sötlimpa"	Norwegian bread "kneipp"	Finnish sweet coffee bread "pitko"	Ginger snaps "pepparkaka"	Danish pastry "wienerbröd"
Ingredients							
White flour	1,000	120	936[a]	500	1,000	1,000	1,000
Brown wheat flour	...	700		500	...	...	...
Rye flour	...	180	...[a)]	...	...	...	450
Water	560	700	490	555	345	...	70
Yeast	35	35	25	33	50	...	70
Salt	17	18	9	13	10	...	9
Sugar/syrup	15-20	...	128 (syrup)	...	210	280/310	70
Fat/margarine	1-2	(+)	...	26	175	385	45 (+1,000)
Nonfat dry milk	2	...	...	...	35	...	...
Egg	...	...	...	...	35	150	180
Flavor/spices	...	...	+	...	Cardamon	Ginger[c]	...
Additives							
Dist. MG/DATA	+	(+)	...	...	...	...	+
Sourdough	...	...	128[b]	(+)	...	...	...
Other	...	...	...	...	Raisins	...	...
Process parameters							
Mixing time, min	10-15	10-20	6-8	5-8	12-20	[d]	2-6
Dough temperature, °C	24-27	24-27	30	26-28	26-28	[d]	12-15
Flour time, min	10-30	10-30	20-40	30	20-40	[d]	5+10+20[e]
Intermediate proof, min	3-15	7-15	3-5	...	...	[d]	50-70
Scale weight, g	400-500	400-800	600-900	900	400-1,000	5-10	30
Final proof time, min	40-60	40-60	30-40	20-35	30-60	...	210-250
Baking time, min	25-30	25-40	20	20-35	20-30	10-15	15-20
Baking temperature, °C	220-240	230	230	230	180-200	190-210	...
Steam during baking	+	+/-	+	+	–	–	–
Product characteristics							
Weight per unit product, g	350-430	400-800	600-800	700-800	400-1,000	5-10	40-60
Specific volume, ml/g	3.5-4.5	2.5-3.5	3.0-3.3	2.5	2.0-2.5	...	...

[a] A mixture of wheat and rye flours (60/40). Note: for total flour (1,000 g) add the 64 g coming from sourdough.
[b] A mixture of rye flour and water (50/50) and sourdough starter fermented overnight.
[c] A mixture of ginger 10, cinnamon 15, and cloves 10 g.
[d] See text.
[e] Time in minutes between three sheeting phases.

mixing, bulk-type relatively short fermentation, and relatively short proofing times. Additives, such as surfactants, are often used to improve dough properties, gas retention, and shelf-life of bread. The short fermentation time is sometimes compensated for by the use of a small amount of sourdough or yeasted pre-ferment in the dough.

The so-called French bread (*franska* in Swedish) serves as an example for white bread processes. Table 6-10 also shows a typical formula for wheat based "mixed" bread, or brown bread. These breads are very popular, and every bakery may make several types of them under fancy names. Another popular bread type in the area, especially in Sweden, is the sweet and slightly sour, brown sweet bread (*sötlimpa* in Swedish) flavored with sugar and syrup. The sourness derives from a relatively small amount of sourdough (Table 6-10).

A specialty product popular mainly in Norway is the *kneipp* (Table 6-10). This is a coarse wheat bread, which arrived in Norway from the German-speaking areas of Europe at the beginning of the century. The original formula used only wholemeal wheat. Norwegian bakers, however, replaced about 25% of the wholemeal wheat with white flour, and today the use of white flour has increased even more in the cities. Nowadays a wide range of kneipp-bread formulas are to be found in Norway, some including sourdough, wholemeal rye, fiber, or other special ingredients.

Sweet coffee bread of various types and sizes are very popular, especially in Finland and Sweden. They are made from white flour, sugar and fat, leavened with yeast, and decorated with egg yolk, sugar and almond flakes. In Table 6-10 a typical Finnish formula for sweet coffee bread (*pitko* in Finnish) is given. Buns of larger size, 400 g or more, are usually made by plaiting three to five dough strips together. Very typical of these products is the characteristic flavor originating from cardamom and often also from cinnamon. Sweet coffee bread and buns are often baked at home, where they are an essential part of the Nordic coffee drinking culture.

Danish pastry is the term used worldwide for the yeast leavened, flaky products made from wheat dough laminated with fat. Two basic versions are generally used in the Nordic countries. The low-fat version is actually called Viennese bread (*wienerbröd* in Swedish), and the high-fat version containing two times more laminated fat is called Danish "wienerbröd," i.e. Danish pastry. Regular white flour may be used for the Danish pastries. However, special flour with no ascorbic acid is also available for this purpose. Very common today is the use of refrigerated and frozen dough techniques in Danish pastry production.

Ginger snaps are an example of typical Northern European cookies flavored with a mixture of cinnamon, ginger and cloves. They are also a popular home-baked product at Christmas time. A typical formula is given in Table 6-10. Syrup, sugar, butter (or margarine) and the spices are brought to boil, then cooled before the rest of the ingredients are mixed. The resulting dough is refrigerated, then sheeted to 2.0–2.5 mm thickness, cut into fancy shapes and baked. Dough for making this kind of ginger snap is also made by bakeries for the consumer market and sold refrigerated.

Future Outlook

Wheat growing in the Nordic countries has increased substantially in the past twenty years. Although rye is still an important bread grain, wheat has the major role. Denmark, Sweden and to some extent Finland are currently self-sufficient with wheat. Only soft wheat and durum wheat is imported to these countries on a regular basis, and strong bread wheat occasionally. Norway in particular is a regular importer of strong North American wheat.

The future development of economic integration in Europe may affect the extent of local wheat production considerably, especially in Finland and Norway, where spring wheat of lower yield is the dominant wheat type. When grown and harvested in good weather Nordic spring wheat may even resemble Northern American spring wheats in terms of gluten properties, and it regularly exceeds the quality of European winter wheats of lower protein content. However, the lower yield and potential sprouting problems make the future of spring wheat growing somewhat uncertain in a situation where competition with imported winter wheats of lower price must be met with lower subsidies.

The possible membership of Norway, Sweden and Finland in the European Community, and the subsequent removal of trade barriers to agricultural products, are challenges for the Nordic food industry, including flour mills and bakeries. The local consumer preferences, high technical level in the flour milling and baking industry, long distances, relatively small markets, as well as limitations in keeping quality of soft bakery goods probably make extensive import of flour and bread unlikely. On the other hand, there is a trend in consumer habits toward a more European-like situation. New technologies, such as frozen doughs, partially baked products, and inert atmosphere packaging also make it easier for imported baked goods to enter the Nordic markets. With future competition in mind, the flour milling and baking industries in Scandinavia have undergone a remarkable structural development over the past few years. We believe that a strong wheat-based local industry and many typical specialties will survive the coming changes in the Nordic countries.

Acknowledgments

The authors are grateful to following persons for valuable information and discussions: Mr. Antero Leino, Finnish Flour Milling Association, Ms. Grete Staerk, The Association of Danish Millers, and Mr. Thomas Robertsson, Swedish Flour Milling Association.

References

Eurostat 1992. Crop production. Quarterly statistics 4/1991. Office for official publications of the European Communities. Bruxelles - Luxembourg.

Fjell, K.M. 1992. Quality control and analytical methods. Pages 75-82 in "Cereal Chemistry and Technology: A Long Past and a Bright Future", 9th International Cereal and Bread Congress, Paris 1992. P. Feillet, ed., Institut de Recherches Technologiques Agroalimentaires des Céréales (IRTAC), Paris.

Finnish Grain Board. 1991. Statistics 1991 (in Finnish). 27 p.

Johansson, H. 1989. A Swedish perspective on plant breeding and baking properties of wheat. Pages 70 - 81 in "Cereal Science and Technology in Sweden", Proc. from an international symposium, June 13-16, 1988, Ystad, Sweden.

Juuti, T. 1988. Targets in the breeding of spring wheat. J. Agric. Sci Finl. 60:281-290.

Mjaerum, J. 1992. Development of quality properties in Norwegian wheat. Pages 181-186 in "Cereals in the Future Diet", Proceedings from 24th Nordic Cereal Congress, Stockholm 1990, H. Johansson, ed., The Swedish Cereal Association.

OECD, 1991. Food Consumption Statistics. Paris.

Perten, H. 1992. The development and international growth of the Falling Number Method. Pages 45-67 in "Cereals in the Future Diet", Proceedings from 24th Nordic Cereal Congress, Stockholm 1990, H. Johansson, ed., The Swedish Cereal Association.

Salovaara, H. 1987. Wheat and flour quality related to baking performance in industrial French bread processes. Acta Agric. Scand. 36:387-398.

Sontag, T., Salovaara, H., Payne, P.I. 1986. The high-molecular-weight glutenin subunit compositions of wheat varieties bred in Finland. J. Agric. Sci Finl. 58:151-156.

Svensson, G. 1987. Situation der Weizenzüchtung in Skandinavien. Getreide Mehl Brot 41:291-293.

Svensson, G. 1992. Improved gluten quality through new breeding technique. Pages 173-179 in "Cereals in the Future Diet", Proceedings from 24th Nordic Cereal Congress, Stockholm 1990, H. Johansson, ed., The Swedish Cereal Association.

Uhlen, A. & Mosleth, E. 1992. The effect of gliadins and high molecular weight (HMW) glutenin subunits on gluten quality of wheat. Pages 163-172 in "Cereals in the Future Diet", Proceedings from 24th Nordic Cereal Congress, Stockholm 1990, H. Johansson, ed., The Swedish Cereal Association.

CHAPTER 7

Wheat Usage in East Asia

Seiichi Nagao

Introduction

Cereals are an important component in the diet of most people in east Asia. In most areas of east Asia, boiled rice rather than wheat flour products is the staple food. Wheat, however, has been traditionally eaten, and some wheat flour foods are peculiar to this area. Each nation in the region has its own eating habit as well as history of wheat usage. Harmonization of the East and West in wheat flour based foods is seen in some areas of east Asia, especially in Japan. A variety of processing methods such as boiling, steaming, frying and baking are utilized in the cooking or manufacturing of wheat flour based foods in this region.

Wheat, Flour, and Flour Usage

Japan

The Japanese have eaten wheat in some form since the 2nd century A.D. Noodles and confectionery products, introduced from China and Europe many years ago, were modified to meet the Japanese taste. The same is true of the wheat produced there. Most flour based foods evolved to a variety of unique foods different from their original forms and tastes. Bread was introduced into Japan by missionaries in the 16th century. It was, however, only in the 1950s that many types of breads and rolls began to be produced in Japan.

Gohan (boiled rice) was the staple food in the Japanese diet well into the 1940s, but since the 1950s a gradual westernization of the lifestyle has been accompanied by a gradual increase in the consumption of flour products. Another factor that has contributed to this increase is the effort by the people in the industries desiring to both introduce wheat flour based foods from overseas and modify them to suit the preference of Japanese consumers. As a result, white pan bread has become a staple food along with boiled rice. The annual per capita consumption of wheat has now leveled off at 31.5 kg, while that of rice has fallen to about 70 kg.

Wheat production began to increase in the late 19th century, reaching a record 1.79 million tons in 1940. Since then, production has fluctuated between 1.78 million tons (in 1961) and 200,000 tons (in 1973). It is currently about 800,000 tons. Almost all the wheat grown in Japan is soft red winter.

The annual wheat consumption in Japan is about 6.3 million tons, about 85% of which is imported. Of the total imports in 1992, 57% was from the United States, 26% from Canada, and 17% from Australia. No. 1 Canadian western red spring (CWRS) wheat is the main starting material for the production of bread flour, whereas Australian standard white (ASW) wheat is indispensable for the production of noodle flour. Each wheat type is, thus, put to appropriate use according to its specific quality characteristics. Because each imported wheat plays an important role in the production of high quality flours, its stable supply is a matter of concern to the milling industry.

The Food Agency, a government organization, has a monopoly on wheat import, while the importing business is executed by trading companies designated by the Agency. Wheat is sold after inspection to each flour mill at a fixed price according to the limit set for that mill. Farmers sell their wheat to the government, which offers them a higher price than that available in the free market. Thus, the mills buy domestic wheat from the government as well.

Although wheat purchase is under government control, flour is traded in a free market system. In 1991, 4.68 million tons of flour was produced. Of that total 95.9% was for industry use and 4.1% for home use. Based on the type of wheat making up the mill mix, flours are customarily classified as strong, semi-strong, medium, and soft (Table 7-1). Within each class, flours are classified by flour grade depending on combination of mill streams used to create them. Thus first, second, third grades, and clear flours exist for all types (Table 7-2). Specialty flours are the blend of selected streams or flours prepared for a certain usage. Table 7-2 (Nagao 1984) shows the approximate protein content, as well as usage of each flour type and grade. Of all the flour consumed in 1991, 36.2% was used in breads and rolls, 35.9% in noodles and pasta, 12.8% in confectionery products.

Among the bread products, square pan bread, confectionery bread, bread for school lunch program and other types accounted for 55.0%, 27.4%, 5.3%, and 12.3%, respectively of production in 1991. Most of square pan bread is white type. Although the shape of white square pan bread is similar to that of pullman type in North America, its texture is peculiar to Japan. The sponge and dough method is used to produce the breads with tender, fine, and moist texture. The consumption of white square pan bread has been on the decline, while that of variety breads (classified as "other" types) has gradually

Table 7-1. Flour Classification and Types of Wheat Milled in Japan[a]

Flour	Wheat
Strong	No. 1 Canada Western Red Spring (Protein: 13.5%)
	U.S. Dark Northern Spring (Protein: 14.0%)
	U.S. Hard Red Winter (Protein: 13.0%)
Semistrong	U.S. Hard Red Winter (Protein: 11.5%)
	U.S. Hard Red Winter (Ordinary)
Medium	Australian Standard White (Western Australia)
	Japanese (Soft red winter type)
Soft	U.S. Western White

[a] ——— = main material, and ······· = supplementary material.

Table 7-2. Average Protein Content of Flour Types/Grades and Their Uses (Protein Content, %)

Types	Grades			
	First	Second	Third	Clear
Strong	Bread (11.5–12.5)	Bread (12.0–13.0)	Gluten/ Starch	Plywood Feed
Semistrong	Bread (11.0–12.0) Chinese noodle (10.5–11.5)	Bread (11.5–12.5)	Gluten/ Starch	Plywood Feed
Medium	Japanese noodle (8.0–9.0) Confectionery (7.5–8.5)	All-purpose (9.0–10.0) Confectionery (8.5–9.5)		Plywood Feed
Soft	Confectionery (6.5–8.0)	Confectionery (8.0–9.0) All-purpose (8.0–9.0)		Plywood Feed

increased. The introduction of a new white pan bread with a softer crumb texture than normal was welcomed by consumers, and resulted in the recovery of total white pan bread consumption. Unique Japanese breads with sweet bean-paste, jam, cream and chocolate-paste fillings, and sweet western goods such as *panettone* and Danish pastry are classified as confectionery breads. White square pan bread is mainly used for school lunch programs, while confectionery bread or "other" types can be chosen as well. Consumption of soft and hard rolls classified as "other" types are also on the rise.

In the pasta and noodle category, wet, boiled, dried and instant noodles, and pasta made up 48.9%, 19.2%, 22.5%, and 9.4%, respectively of 1991's total production. Confectionery products are classified as baked goods (hard biscuit, cookie, cracker, *bolo*, *senbei*, and wafer), fried goods (doughnut and fried dough cookie), fresh Japanese style goods (baked bun and steamed bun), and fresh western style goods (*castilla* and cake) (Nagao 1981). Other minor uses of flour include *tempura* and *yaki-fu*.

Korea

Wheat consumption in South Korea in recent years averaged 2.0 to 2.2 million tons/year. However, the principal source of dietary carbohydrate in the Korean diet remains rice. Domestic wheat production is small. More than 80% of the imported wheat comes from the United States, but the percentage imported from Australia has been gradually increasing. Of the imported wheat, that of low protein content (U.S. western white [WW], Australian soft and ASW) comprises about 55% of the total. Hard red winter (HRW) and dark northern spring (DNS) wheat from the United States are the principal hard wheats.

Noodles account for about 40% of total flour use, breads and pastries for 10 to 15%, and confectionery products for 15 to 20%. Only 3% of flour produced is for home use. WW and ASW wheat are the main starting materials for noodle flours. The protein content of noodle flours ranges from a low of 8% to a high 11%. The protein content requirements for bread and pastry flours are 12.5–13.5% and 11.0–12.0%, respectively. Cakes, biscuits, crackers and snacks are made from a soft wheat flour of 7.0–8.5% protein. Roughly 18% of the flour produced is all-purpose type (9–10% protein) which is milled from a blend of hard and soft wheats. This is the primary home use flour, as well as the flour for specialty foods, such as dumpling.

Unique and popular wheat flour foods in Korea are *Naeng-myon* (cold noodle), steamed bun and steamed cake. Soft wheat flour of 7.5–8.0% protein is used for making a nonalcoholic beverage called *Makoll*.

Taiwan

Imports from United States and Canada supply Taiwan's annual wheat consumption of about 0.9 million tons. Of total flour consumption, about 30% goes to noodles including raw, boiled noodle and thin types (*Nian Xian*), 6% to instant dried noodles, 21% to breads and rolls, 17% to traditional Chinese products including dumpling, bun, steamed bread, puff pastry and twisted cruller (fried noodle stick), 10% to confectionery products including cake, cookie, cracker, biscuit and pastry, and 9% to vital wheat gluten products.

The ash content of most flours marketed in Taiwan is 0.40–0.45%. A premium grade flour (0.36 to 0.40% ash) of 10.5–11.5% protein is used for making wet raw noodles. Boiled alkaline noodles are made from flour of 11.3–11.5% protein, which is milled from DNS, HRW and CWRS wheat. A flour of 11.5 to 13.0% protein and 0.45 to 0.55% ash content is processed to instant dried noodle. WW wheat is often blended with hard wheat in the milling of this flour. Pan bread flour is of 12.8–13.0% protein content. Traditional Chinese products are made from a lower protein flour (10.5 to 11.5%), which is milled from a blend of hard and soft wheat. Flour for cake and pastry making is 7.0–8.0% protein, and milled from WW wheat.

The People's Republic of China

Wheat production in China is nearly 100 million tons a year. Nevertheless, 135–159 million tons of wheat and flour were imported per year in 1988–1990. The United States, Canada and Australia are China's main wheat suppliers. Although rice is still the major staple in China as a whole, more than one-third of the Chinese population consumes wheat as its staple food. It is estimated that 50–60% of total flour consumption is as noodles, 30 to 40% as steamed breads including buns with filling and Chinese-style appetizers (*Dim sum*), and less than 10% as pastries.

A flour high in ash (>0.75%) and 10.5–12.5% protein is used for making noodles and steamed breads. Some flour mills in urban areas produce a flour of better quality (0.70% or slightly less in ash) and 11% or more in protein content, which is mainly used for production of western style breads and pastries. Cake flours are 0.60–0.70% in ash and <10.5% in protein. A small amount of special grade flour (<0.60% ash, >11.5% protein) is produced by a handful of flour mills to meet specific customers' requirements.

The Philippines

Wheat production is almost nonexistent in the Philippines. Total imports of wheat and flour were 1.2–1.6 million tons over the period 1988–1990. The United States is the major wheat supplier and about 70% of the wheat consumed in the Philippines is hard wheat. Flour imports are on the decline in recent years, due to the increase in the capacity of domestic flour mills.

Approximately 20% of all flour produced goes to make local breads such as *pan de sal*, 18% is used for loaf bread, 16% for buns and rolls, 3% for steamed bread, 8% for instant noodle, 6% for Chinese noodle, and the balance to produce cookie, cracker, cake, and for home use. The protein content of flours used in *pan de sal* and loaf bread is 13%. Buns, roll and Chinese noodles require a flour of 12–13% protein. Instant noodles are made from a flour of ~10% protein. Cake flour, milled from WW wheat, is generally 7–8% in protein content.

Indonesia

Yearly imports of wheat and flour by Indonesia are 1.6 to 1.9 million tons. No wheat is produced by the country. Of the wheat imported, ASW accounts for the major portion and is used mainly for the production of Chinese type wet noodles. The balance of the imports were from the United States and Canada.

Chinese type wet and instant noodles account for about 45% of flour use, bread and cake for 30%, and cookie and cracker for 20%. The protein contents of flours for instant noodle and Chinese type wet noodle are 11 to 12% and 10 to 11%, respectively. Bread flour is not high in protein content (11 to 12.5%).

Singapore

Singapore imported 330,000 to 440,000 tons of wheat and flour per year in 1988 to 1990. Of the wheat imported, ASW used for making Chinese type noodle accounts for the major portion. A small amount of DNS and WW wheat is imported from the United States. About 40% of the flour marketed in

Singapore goes to Chinese type wet and instant noodles, 40% to bread and cake, and 15% to cookie and cracker.

A flour of high protein content (12–14%) is used for bread making. Chinese type wet noodle is made from a flour of 10 to 11% protein content.

Malaysia

Wheat and flour imports from 1988 to 1990 were 650,000 to 840,000 tons per year. Malaysia does not produce wheat in commercial quantities. Of the wheat imported, ASW dominates the market.

It is estimated that about 45% of total flour consumption is as Chinese type wet and instant noodles, 30% as bread and cake, and 20% as cookies and crackers. The protein content of flour for Chinese type wet noodle is 10 to 11%, which is milled mainly from ASW wheat. Hard wheat blended with some ASW is milled to make a bread flour of 12 to 14% protein content. A small amount of Australian soft wheat is imported to mill a flour for cake and biscuit production.

Thailand

Wheat and flour imports increased from 290,000 tons in 1988 to 370,000 in 1990. No wheat is produced in commercial quantities. The principal supplier of wheat is the United States, followed by Australia and Canada.

About 22% of the flour marketed is used for the production of bread, 21% for biscuit, 21% for all-purpose uses including cookie, bun, pizza crust, snack and other confectionery products, 13.0% for noodle, and 18% for shrimp feed. A flour for bread is 0.50 to 0.55% in ash and 13.0 to 14.0% in protein content, which is milled mainly from high protein DNS wheat. WW, ASW and HRW (low protein) wheat are blended to mill all-purpose flour, which is 0.40 to 0.52% in ash and 8.5 to 11.0% in protein content. A premium grade flour (0.36 to 0.40% ash) of about 8% protein is used for making cake. The wheat blend used for milling cake flour is almost similar to that for all-purpose flour. However, classification is often used to lower the protein content of a cake flour.

Baked Products from Wheat Flour

Breads and Rolls. Most kinds of Western breads and rolls are produced in Japan. Although some are very similar to those in the West, others have been modified to meet local preferences. As previously stated, most bread is made by the sponge and dough method, but the crumb texture and grain of white square pan breads are characteristic of Japan. They have tender, fine and moist crumb texture. In most east Asian countries other than Japan, Western breads and rolls are becoming increasingly popular. Their quality, however, differs from one country to another depending on the available wheat, yeast and production techniques.

A number of unique local breads are produced as well. They include confectionery breads with sweet bean-paste, jam, cream and chocolate-paste fillings in Japan, and *pan de sal* in the Philippines.

Chinese Flat Bread. *Lao Bing* is a Chinese style flat bread requiring a flour of 8.5–10% protein. To make a typical *Lao Bing*; flour, salt, and lukewarm water ~30°C are mixed (100:2:10) into dough. The dough, covered with a wet cloth, is rested at room temperature, cut into pieces, rounded, and sheeted to 13–17 cm wide by 23–26 cm long. Peanut oil is brushed on its surface. Then, it is folded, rounded, sheeted again, and is baked in a pan, turning for several times until both sides become golden brown.

Sponge Cakes. Sponge type cake is popular in Japan as it is in other nations in east Asia. The basic baking techniques are similar to those in the West. Two formulas are typical (Akutu et al 1963; Nakae 1975). The first consists of flour, eggs, sugar, and milk and water (100:100:100:30–35). The second increases the egg to 170–200, the sugar to 120–160, and replaces butter for milk, (30 to 50). Emulsifier and foaming agents containing sucrose fatty acid esters are often used.

Flour for Japanese sponge cake ranges from 7.0–7.8% in protein and 0.33–0.38% in ash content. It should possess what has been referred to as "mellow" gluten. A short patent flour from WW wheat suits these requirements. High amylase activity is undesirable in sponge cake flour, as it causes collapse of the final product (Nagao 1975). The sponge cake baking quality of white club (WC) wheat is generally superior to that of soft white (SW). Thus, the more WC is blended into WW wheat which is the mixture of WC and SW, the more favorable it is for processing (Nagao et al 1977).

Sponge cake formulas may be used to bake layer cake and short-cake, as well as fancy cake. The baking technique for sponge cake has been also applied to the automatic production of roll cake, made by rolling up a long cake sheet ~1 cm thick.

Castilla. *Castilla*, named for a kingdom in Spain, was introduced into Japan by the Portuguese in the 16th century. Modifications in the original formula, baking process and flavor have made it a Japanese cake.

Ingredients for *Castilla* include: flour (100%), eggs (200–210%), sugar (180–210%), starch syrup (15–30%), water (10–20%), and sweet sake (*Mirin*) (Sakurai 1959; Niimi 1965; Tokyo Koto Seika Gakko 1971). The whole eggs are beaten, sugar added, and the mixture blended at low speed, after which remaining ingredients are then added gently. The batter is poured into a wooden frame placed on a steel plate covered with a sheet of craft paper, and the frame is covered with a second heat-intercepting plate. During the initial baking (25 min at 180°C) the upper plate is removed. A second wooden frame is placed on the first (Fig. 7-1), and the temperature is reduced to 150°C. Total baking time is 55 to 60 minutes.

Short patent flour milled from WW or low-protein Japanese wheat has been favored for *Castilla* production. Quality requirements for the flour are similar to those for sponge cake flour. *Castilla* and sponge cake formulations differ in the higher proportion of sugar, eggs, honey, and/or starch syrup in the former. The sugar and honey and/or starch syrup aid in maintaining the cake's tenderness and moistness. It is this

Figure 7-1. Wooden frames for *castilla* baking.

soft, moist masticatory sensation which is a most important characteristic of *Castilla*.

Baked Japanese Buns. *Manju*, the generic term for Japanese buns, are classified either as lightly baked or steamed. *Kuri-manju* (chestnut bun with sweet bean-paste filling) is a typical baked product. A short patent flour (0.33–0.40% ash, 7.5–9.0% protein) from a blend of WW and Japanese wheat is used for *kuri-manju*.

The *kuri-manju* bun formula contains (Niimi 1965; Kajitani 1968) flour (100%), eggs (35–55%), sugar (50–60%), starch syrup and honey (5–15%) (optional), and baking powder (sodium bicarbonate, ammonium chloride, cream of tartar, and baking aluminum phosphate) (0.5–1.0%).

The dough is formed by beating the eggs lightly, adding the sugar, starch syrup, honey and baking powder, and mixing. After adding the sifted flour, the mixture is mixed at low speed. The filling is prepared by heating bean-paste with sugar and starch syrup. Chestnuts (each chopped in eight pieces) are added at the final step.

In most large confectioneries, co-extruders are used to fill the dough sheet and to mold it into the desired shape. The continuous, automatic process makes possible the production of a uniform, high quality product at lower unit cost, that meets sanitation requirements. The operation is performed manually in small confectioneries. Egg yolk mixed with sweet *sake (Mirin)* is brushed on the surface as it exits from the machine or after hand molding, in order to add a glaze to the product. Baking is at 180°C.

Because a soft and tender texture is desirable, care is taken at the mixing stage in the baked product. Overmixing the dough results in an objectionable tough product. The external appearance is a very important quality factor, too.

Baked Chinese Buns. *Yit bien* (moon cake), a baked bun with filling, is popular in China and other nations with a significant Chinese population. It is served at the Festival of the Moon of the Chinese calendar (~August 15). The crust of the bun is decorated with embossed figures of rabbits, frogs, pagodas, etc. (Fig. 7-2).

A typical *yit bien* dough formula contains (Niimi 1965; Tokyo Koto Seika Gakko 1971) flour (100%), sugar (50–55%), starch syrup (25%), lard (35–40%), sodium bicarbonate (1%), aluminum phosphate (0.04%), and water (15–25%). The filling is quite characteristic and consists of chopped almond, cashew, walnut, and watermelon seeds, sugar, white sesame seeds, steamed flour (prepared by steaming flour for about 30 minutes), powdered rice, lard, and Chinese watermelon cooked with sugar. All ingredients other than flour and rice are combined and heated. The two flours are added during the final cooking stage.

The dough is prepared by mixing sugar, starch syrup, aluminum phosphate and water, adding liquefied lard and mixing again. Soda dissolved in water is then added. Finally, sifted flour is added and the ingredients are mixed manually at low speed to a fairly firm dough. The dough, rested for about an hour, is divided into small balls. These are flattened, the filling is placed in the cavity, sealed by folding the edges. The filled piece is molded in a wooden embossing form, egg yolk mixed with sweet wine or caramel solution is brushed on the surface of the dough, and it is baked at 190° for about 15 minutes.

Short patent all-purpose flour milled from soft wheat or a mixture of 80% soft and 20% hard wheat flour is suitable for *yit bien*. Product quality is determined by the clarity of the pattern embossed on the surface. The baked product should have a smooth, fine texture that tends to melt in the mouth.

Biscuits and Cookies. Biscuits are baked in most countries in east Asia. In Japan, the products are classified as either hard or soft types. Those soft biscuits, made by a richer formula than usual, are termed cookies. Formulas and manufacturing processes for biscuits and cookies are similar to

Figure 7-2. *Yit biens* (moon cakes).

those employed in the West (see Chapter 1). Short patent, soft wheat flours of low protein content, fine granulation, and mellow gluten quality are suitable for soft biscuits. This is similar to North American "cookie flour." Occasionally first grade all-purpose flour may be used for special products. For hard biscuits, first or second grade all-purpose flour of moderate protein content (8–9%) is used.

Chinese Cookies. A variety of Chinese cookies are found in east Asia. *Hsing jen ping* (Chinese almond cookie) is typical of this class of products. A formula for almond cookie includes flour (100%), sugar (50–60%), eggs (10–30%), lard (60–80%), salt (0–0.5%), sodium bicarbonate (0.02–0.5%), halved almonds (20–40%), and flavorants (to taste). Lard and sugar previously creamed to prevent the lard from separating in the finished product are mixed with eggs, flavorants, and sodium bicarbonate. Sifted flour is added at low speed or by hand to produce a fairly firm dough. The dough is formed into small balls and placed on a baking pan. A cavity is formed in the center of each piece with a dowel and filled with almond halves. Baking is for 10 minutes at 180 to 190°C. The dough may be brushed with egg yolk prior to baking to form a glaze.

All-purpose soft wheat flour is usually used for Chinese cookies. Flour quality requirements are similar to those for Western cookies.

Crackers. The crackers produced in Japan resemble the American type. Their formulas and manufacturing processes are almost identical to those found in the United States (see Chapter 1).

Senbei, the generic name for Japanese type crackers are differentiated into wheat flour cracknels and rice crackers. *Kawara-senbei* (tile-shaped cracknel) is most popular among a variety of wheat flour cracknels. Its typical formula includes flour (100%), sugar (110–130%), eggs (10–20%), sodium bicarbonate (0.03–0.1%), and water (60–80%). The sugar and eggs are whipped, and the sodium bicarbonate (dissolved in a little water) is added with the flour and mixed well. The remaining water is added and mixed to create a uniform batter. This batter is poured into baking molds that are heated over a fire, the molds being inverted carefully midway in the heating.

Soft wheat flours with 0.35 to 0.55% ash and 7 to 10% protein content are used for making *senbei*. A crispness characteristic of the product is a primary quality requirement. The product should soften readily in the mouth without becoming gummy.

Yaki-fu. *Yaki-fu*, a traditional puffed food baked from wheat gluten supplemented flour, has been a part of the Japanese cuisine for many years. Gluten extracted from a hard wheat flour of high protein content is kneaded with a high extraction hard wheat flour in a special vertical mixer. The ratio of gluten and flour is 1:3 to 1:1 although this varies according to the type of product being produced. A small amount of potassium bitartrate, sodium bicarbonate and aluminum phosphate are added to flour prior to mixing (Endo 1980). After they are mixed well, the dough is kneaded in a special kneader (a small compact kneader with blades) to complete the dough development. After resting at room temperature, dough is divided, molded into the shape for final products, and is puffed in a special, puffing oven.

Although originally made at home, modernization in processing methods and apparatuses have made it possible to produce a variety of quality *yaki-fu* in characteristic shapes and flavors. By using frozen gluten, manufacturers can bake quality *yaki-fu* at any time.

Other Baked Products. Wafers are favored by the very young, and are frequently served with ice cream. Their formula, mixing and baking processes are similar to those in the West. First grade or a blend of first and second grade flour from WW wheat is preferred for the production. **Cones**, or containers for ice cream, are classified as sugar cones and ordinary cones. Like wafers, their formulas, production, product characteristics and flour requirements are similar to those in the West. They are baked in cup, bar, and plate shapes.

Butter cakes including pound cake, cup cake, *madeleine*, and lemon cake are popular in Japan. Their formulas, mixing and baking procedures are similar to those in the West. Flour quality requirements are similar to those of sponge cakes.

Hot cake (pan cake) is made both at home and at restaurants which serve them with a variety of toppings. Hot cake mixes are available in Japan for both home and restaurant use. A typical formula for hot cake batter is similar to Western formula and includes flour (100%), sugar (30–40%), eggs (70–80%), milk (40–50%), butter (0–5%), and baking powder (3–4%). First grade soft wheat flour is most suitable for hot cakes.

Japanese waffles with jam, cream or chocolate filling are favored by many as a snack. Their formula is sweeter than that of hot cakes and includes flour (100%), sugar (100–110%), eggs (100–120%), starch syrup (10–20%), baking powder (1–2%), water (30–40%), and flavoring. Mixing and baking processes are similar to those in the West. Filling is spread on the baked waffle, which is then folded. Short patent flour from WW wheat is the most suitable starting material.

Noodles and Pasta

Japanese Noodles. Japanese noodles are creamy white in appearance and soft in texture. They are classified by width into very thin (*so-men*), thin (*hiya-mugi*), standard (*udon*), and flat (*hira-men*). Cooked *so-men* or *hiya-mugi*, served cool, is

Figure 7-3. *Udon* served in hot soy sauce soup with oysters, shell-fish, *tofu* and vegetables.

eaten in summer by dipping in cool soy sauce soup. Cooked *udon* or *hira-men* is served in hot soy sauce soup with a boiled egg, *tempura*, fried *tofu*, boiled fish paste, vegetables, etc. in the cool seasons (Fig. 7-3). *Udon* or *hira-men* served with a special sauce is becoming quite popular among young people.

Boiled noodle (*yude-men*) is a popular form of *udon*. It is sold either unpacked, packed in a polyethylene film, or packed tightly for extended shelf life. Dried noodle (*Kan-men*) is a storable form produced by controlled drying of uncooked wet noodle strands. A new type of dried *udon* requiring less cooking time is shown in Figure 7-4. Its cooking time, 3 to 4 minutes, is less than half of that of ordinary noodle. However, the palatability of the cooked product is similar to that of the regular type. Seen in cross section, the noodle strand has two depressions which close during boiling.

Although most noodle manufacture is mechanized, two types of handmade noodles are available. In the manufacture of handmade very thin noodle (*tenobe so-men*), soft dough cut into strings is twisted and stretched repeatedly to create the final product. Handmade standard or flat noodle (*teuchi udon* or *teuchi hira-men*) is sheeted and cut rather than stretched. The basic process for automated noodle making requires mixing (blending) the raw materials, sheeting the dough, combin-

Figure 7-4. Dried *udon* for short time cooking.

ing two sheets, rolling, and cutting the sheet into strands. The formula for Japanese noodle includes flour (100%), salt (2–3%), and water (28–45%). The quantity of salt is adjusted based on the type of noodle, market requirements, and climate. Dried noodle requires more salt (2 to 3%) than does boiled noodle (2%). In winter, less salt is required.

A variety of mixers differing in type and size are used in the industry. Although blending of flour and water is the main function of noodle mixers, some are designed to knead the dough as well. The most common practice is to mix for about 10 to 15 minutes in a horizontal mixer to produce a stiff, crumbly dough.

Most noodles are made from flours of 0.36 to 0.40% ash and 8 to 10% protein. A dull white hue to the product is not desirable, while a bright creamy tone is considered acceptable. Good surface appearance, favorable texture, minimal cooking loss and high noodle yield are important noodle quality factors associated with flour. Extremely low protein content, abnormally weak gluten and low amylograph viscosity cause problems in noodle manufacture and reduce acceptability of the finished products.

Chinese Type Noodle. Chinese type noodle (Fig. 7-5), which is called *ra-men* in Japan, is eaten in most countries of

Figure 7-5. Chinese type noodle served cool.

east Asia. Their quality varies. *Ra-men* is made from hard wheat flour (100%), water (32–35%), *kansui*, a mixture of sodium carbonate, potassium carbonate and sodium phosphate (~1%), and salt (0–2%). *Kansui* reacts with flour components during processing to produce noodle with characteristic color, flavor and texture. The uncooked wet product is sold to noodle restaurants and retail shops.

In Japan, the presence of a uniform, light yellow color in the noodle is of high concern to consumers as well as for noodle manufacturers. Any visible specks or off color in the noodle is not desirable. High protein (10.5–12%) but mellow gluten quality is required of the flours for use in noodles. Top grade flour streams of a specially selected wheat blend which satisfy these requirements are carefully chosen and combined to make quality flours of very low extraction (0.33–0.38% ash content).

Steamed Chinese type noodles sautéed on a hot iron plate with sliced vegetables and pork is popular among the young in Japan.

Naengmyon, a Korean cold noodle, is made from 100% all-purpose flour (9.5–10.5% protein); 10–30% buckwheat flour; 80–85% water; and 2% alkaline water consisting of 40 parts sodium carbonate, 5 parts sodium bicarbonate, and 55 parts water. Up to 20% starch may be added to this formula. The product is sold both fresh (wet) and dry. In the manufacture of wet product, drying is omitted

Instant Noodle. The first instant noodle called *chicken ra-men* appeared in Japan in 1958. The introduction in 1971 of instant noodles packed in a cup or a bowl was a major advance owing to its convenience (Shibata 1990). Instant noodles packed in polyethylene bags are either fried or non-fried type. Plain fried noodle and cup noodle include not only Chinese but also Japanese noodles.

About 0.2 parts of *kansui* powder is added to 100 parts of flour in the production of instant Chinese type noodle. Though their basic production is similar to those of ordinary Chinese type noodles, noodle strands for cup and non-fried instant type are cut a little thinner to allow for rapid hydration. In the production of instant noodles, noodle strands are waved and steamed in a tunnel steamer for 1 to 3 minutes at 95 to 100°C. In the case of the non-fried type, enough steaming is necessary to complete the gelatinization of starch. Drying by frying is done for 2 to 3 minutes at 135 to 140°C. Hot blast frying is usually done for more than 30 minutes at 80°C.

Buckwheat Noodle. Buckwheat noodle is light brown or grey in color and unique in texture and flavor. It is made from buckwheat flour, hard wheat flour, and water. The ratio of hard wheat flour to buckwheat flour varies according to the type of product.

Pasta. Spaghetti and macaroni are favored by the young in Japan. They are becoming popular in most other countries of east Asia, where imported pasta dominates the market. In Japan, pasta is produced from 100% durum semolina milled from Canadian or U.S. durum wheat. The amount imported to Japan is not large. A variety of Italian and Japanese pasta sauces specifically developed by the industry have contributed to the increase in pasta consumption.

Steamed Breads and Buns

Steamed Bread. Steamed Chinese bread is eaten in most countries of east Asia. There are three types of products: those without any filling, with sweet bean-paste filling, and with meat filling (Fig. 7-6).

Typically, a dough is prepared with flour (100%), fresh yeast (2–4%), sugar (5–10%), salt (0–0.5%), lard (3–5%), and milk and warm water (50–60%). Active yeast with sugar dissolved in lukewarm water is added to the sifted flour and mixed to a

Figure 7-6. Chinese steamed breads with meat filling (left) and sweet bean-paste filling (right).

stiff homogenous dough. Salt and liquified lard are added during mixing. The dough is fermented to double in bulk and remixed. It is then shaped into a cylinder about 5 cm in diameter and sliced into pieces of 2 to 3 cm wide. Each piece is shaped, filled, and sealed. The rounded dough, on a square of parchment paper, is placed on a rack, permitted to rise for 7 to 10 minutes, then steamed for about 10 minutes.

The quality of flours for Chinese steamed bread varies widely. First grade soft wheat flour is generally desired to give the dough softness and tenderness and impart a fine and silky texture to the final product. However, if the flour is excessively weak or low in protein content, blending of hard wheat flour (usually 1 part to 2 parts soft wheat flour) is employed. The surface of a steamed bread must remain white and glossy without turning yellow after steaming. A high quality steamed bread should be light and not gummy or sticky in the mouth. The space between dough and filling should be minimal. Steamed products can be frozen, packed, and delivered to bakery shops or convenience stores, where they are thawed and rewarmed in a special steamer to be sold to customers.

Steamed Bun. Steamed Japanese buns are of two types: *saka-manju*, prepared by steaming a fermented dough with sweet bean paste filling, and *mushi-manju*, a similar product but made from chemically leavened dough. The leavening agent for *saka-manju* may be a fungus from a sweet fermented rice drink (the lees after *sake* fermentation) or fresh yeast.

Mushi-manju bun is made from first grade confectionery or all-purpose flour. Bright white color and soft gluten quality are desired in the flour. Flour protein requirements vary according to the type of product. In this case soft wheat flour of 7.0–8.5% protein is suitable. On the other hand, first grade all-purpose flour is preferred for *saka-manju* manufacture. Whiteness combined with moderate protein content (8–9%) is required. For both products, quality is distinguished by a soft and tender crust and a shinny white surface.

Chinese Steamed Dumpling. A hard wheat flour of 10 to 12% protein and <0.5% ash is used for steamed dumpling making. Typically, a dough is prepared with flour (100%), fresh yeast (2%), and water (~60%). Fresh yeast dissolved in lukewarm water, sifted flour and water are mixed to a stiff homogeneous dough, and fermented for about half an hour. The dough, shaped into a cylinder, is cut into small (~15 g) pieces, sheeted to ~2 cm, filled and wrapped (scaled). Filled dumpling are steamed for ~6 min.

Fried Products from Wheat Flour

Tempura. *Tempura*, a delicious Japanese food, is served in tempura restaurants that feature individual variations in preparation and presentation (Fig. 7-7). It is also prepared at home.

A *tempura* batter formula includes flour (100%), eggs (25%), and water (135%). The eggs are lightly beaten, the water added, and the light beating continued. Finally, the flour is sprinkled in with light mixing to form a smooth batter. Over-mixing results in an undesirable sticky batter. WW wheat is used for milling *tempura* flour. Low protein content and soft gluten quality are desired so as to produce a crisp, tender and thin fried coating.

Shrimp or vegetables are dipped in the batter, and the coated material is fried in high quality vegetable oil at 160 to 190°C. Cooking is complete when bubbles generated by the frying become small and few. The batter coating plays an important role in protecting the material from the high oil temperature and in preventing excessive moisture evaporation. It thus aids in maintaining food flavor and even enhances palatability through the absorption of a little oil.

Prepared *tempura* batter mixes are available, most of which

Figure 7-7. Tempura.

are for family use. A quality mix which contains wheat flour, wheat starch, egg yolk powder, baking powder, salt, seasonings and enzymes is used by adding water (3 parts water to 4 parts mix).

Doughnut. A clear flour of <0.7% ash and 13–15% protein is commonly used to produce *You Tiao*, a Chinese fried doughnut stick. A typical formula includes flour (100%), ammonium bicarbonate (1.5%), baking soda (0.9–1.5%), ammonium aluminum phosphate (1.5%), salt (1.22%), and water (~62%).

The dough is prepared by mixing flour, ammonium aluminum phosphate dissolved in 25% of formula water, and ammonium bicarbonate, baking soda and salt dissolved in the remaining water. Mixing is done in three stages for 30, 20 and 10 seconds, respectively. After the first and second mixing, the dough is rested for 20 minutes. After the third mixing, it is divided into 50-g pieces, made up into oblong shapes, wrapped with a PVC sheet or cloth greased with shortening, and rested for several hours. Each 50-g dough piece is fried in oil at 200 to 250°C.

The consumption of Western type doughnuts has been increasing in most east Asian countries. In Japan, a variety of doughnut mixes, for both yeast-raised and chemically leavened products, are commercially available. Flour quality requirements, formula, processing techniques, and equipment employed are similar to those in the United States (Chapter 1).

Fried Sesame Ball. A medium protein semi-hard wheat flour mixed with a soft wheat flour is used for the manufacture of fried sesame ball in Taiwan (Fig. 7-8). Typically, its formula includes flour (100%), sugar (30%), black sesame seeds (8%), *tofu* (40%), and eggs (20%).

A hole is made in the center of a mound of sifted flour, other ingredients are put in hole, and mixed and kneaded into a dough by a hand. The dough is rested for 15 to 20 minutes, rolled into a flat sheet using a noodle making machine, cut into 2 cm × 5 cm pieces, and fried in oil at 185°C.

Karintou. *Karintou* (fried dough cookie) is a Japanese confection originally imported from China. During the 18th century, it was modified to differ from the original. *Karintou* can be eaten soft or hard. Soft *karintous* are produced by yeast fermentation of a hard wheat flour dough to result in a light and airy finished product. Hard *karintous*, on the other hand, are made from a chemically leavened soft wheat flour dough, which processes into a coarse and slightly hard product.

Figure 7-8. Fried sesame ball.

A typical formulation for hard *karintou* consists of 0–3% sugar, 0.3% skim milk, 0–1% salt, 4–5% sodium bicarbonate/ammonium carbonate, and 45–60% water based on 100% flour. The ingredients are mixed, sheeted, cut, fried, coated with sugar syrup, and cooled.

Industrial Uses of Wheat

About 20% of the clear flour produced in Japan is used as an ingredient in the adhesive agents employed in the plywood industry. The flour is mixed with resins to make an adhesive paste. The proportion of flour in the resins varies from 20 to 300% by weight based on 100% resin, depending on the type of plywood to be made. Flour serves as a filler, as a means of viscosity control, and for moderating excessive adhesion strength of the resin. Flour for use in adhesives is usually 2–2.5% ash, and of a particle size to pass through a 100-mesh screen. Coarse bran particles cause poor adhesive action by clogging roller machine surfaces. The clear flour from hard wheat milling differs from that from soft wheat in its effect on adhesive viscosity. Some plywood manufacturers in the Philippines use a long patent flour from soft wheat milling.

Acknowledgments

The author is very grateful to R. Bratland, H. Terada (Tokyo), E. B. Yoon (Seoul), K. H. Lu (Taipei), M. Weimar (Beijing), R. Chung (Singapore), Boy C. NG (Manila), all members of U.S. Wheat Associates; A. Hunter, Y. Takahashi (Australian Wheat Board, Tokyo), T. Ouno (Canadian Wheat Board, Tokyo), and H. C. Hsu (China Grain Products Research and Development Institute, Taiwan) for information on wheat uses in east Asian countries other than Japan; and to O. Shoda and H. Otsubo (Nisshin Flour Milling Co., Tokyo), for their approval of this contribution to the monograph.

References

Akutu, S., Freundlief, H. and Matsumoto, H. 1963. Bakery Handbook. (In Japanese) Fujisawa Mfg. Co.,Ltd., Osaka, Japan.

Endo, E. 1980. Wheat Protein - Its Chemistry and Processing Technology. (In Japanese) Shokuhin-kensyusha, Tokyo.

Endo, S., Karibe, S., Okada, K. and Nagao, S. 1988. Factors affecting gelatinization properties of wheat starch. (In Japanese) Nippon Shokuhin Kogyo Gakkaishi 35: 7-14.

Endo, S., Karibe, S., Okada, K. and Nagao, S. 1988. Comparative studies on the quality characteristics for prime starch and starch tailing. (In Japanese) Nippon Shokuhin Kogyo Gakkaishi 35:813-822.

Kajitani, K. 1968. Mass-productive confectionery by machine. (In Japanese) Food Machinery Trade Center, Nagoya, Japan.

Nagao, S. 1975. Why does low amylograph viscosity cause problem? (In Japanese) Grain Importers Assoc. J. 323:35.

Nagao, S., Ishibashi, S., Imai, S., Sato, T., Kanbe, T.,Kaneko, Y., and Otsubo, H. 1977. Quality characteristics of soft wheats and their utilization in Japan. II. Evaluation of wheats from the United States, Australia, France, and Japan. Cereal Chem. 54:198-204.

Nagao, S. 1981. Soft wheat: Production, Breeding, Milling and Uses ed. by Yamazaki, Y. and Greenwood, C.T. AACC 269, St. Paul, MN.

Nagao, S. 1984. Wheat and Its processing. (In Japanese) Kenpakusha Ltd., Tokyo.

Nakae, K. 1975. Bakers index. (In Japanese) Shokuken Center, Ltd., Tokyo.

Niimi, K. 1965. Encyclopedia of Manufacturing Methods for Japanese Confections. (In Japanese) Numata Shoten, Ltd., Tokyo.

Sakurai, Y. 1959. Encyclopedia of Foods. (In Japanese) Dobun Shoin, Tokyo.

Shibata, S. 1990. Knowledge of wheat flour foods ed. by Shibata, S. and Nakae, T. (In Japanese) Saiwai Shobou Ltd., Tokyo.

Tokyo Koto Seika Gakko. 1971. Textbook of Japanese, European, and Chinese Style Confections. (In Japanese) Numata Shoten, Ltd., Tokyo.

CHAPTER 8

Wheat Usage in the Indian Subcontinent

Jiwan S. Sidhu

Introduction

The Indian subcontinent (India, Pakistan, Bangladesh, Nepal and Sri Lanka) has a total area of 4,330,210 square kilometers: India, 3,179,476 sq. km.; Pakistan, 803,943 sq. km.; Bangladesh, 142,776 sq. km.; Nepal, 139,166 sq. km; and Sri Lanka, 64,849 sq. km. As per the 1990 census the total population was 1,126 million, of which 705 million were engaged in agriculture (Anon. 1990).

The climate and soil in the Northern part of India and Pakistan are suitable for wheat cultivation. In the subcontinent, India leads in wheat production followed by Pakistan. Among those countries, India is nearly self-sufficient in meeting its wheat requirements except in years of severe drought. On some occasions India does export wheat to other deficient countries of the region. Pakistan grows ~15 million tons of wheat annually, but is not yet self sufficient. Bangladesh's annual wheat production is about a million tons, an amount insufficient to meet demands. Nepal grows only 766,000 tons of wheat annually and, like Pakistan and Bangladesh, imports the remainder. Sri Lanka does not grow any wheat and all of the requirements are met by imports.

Prior to the mid-sixties, the entire region grew only tall cultivars with a single dwarfing gene. Those wheat cultivars were susceptible to lodging when a higher dosage of nitrogenous fertilizers was applied. Because of this, the grain yields were extremely low. In the mid-1960s double dwarf wheat cultivars were introduced which were both high yielding and tolerant to lodging under high doses of nitrogenous fertilizers, irrigation and other agronomic practices. That led to a tremendous growth in wheat production in this region and this was commonly known as the Green Revolution.

The majority of the population in India and Pakistan are wheat consuming. In Sri Lanka, Nepal and Bangladesh the population consumes mainly rice. However, wheat plays a vital role in meeting the biological needs of a large segment of the population of the latter region. Over 80 percent of the wheat produced is milled into whole wheat meal by stone/disk mills (*Atta chakkies*). Keeping aside about 10% for seed and other miscellaneous purposes, the remaining is milled by roller flour mills into white flour, semolina and red dog. Most of the wheat produced in the Indian subcontinent is of medium strength and is quite suitable for the preparation of indigenous baked products like *chapaties, poories, paronthas*, etc. Apart from these products, a sizable quantity of wheat is also used in the preparation of various sweet and savory snacks. A small amount of wheat also finds use in the preparation of Western style baked products like bread, biscuits, cookies, crackers and cakes. Although the wheat cultivars grown in this region are of medium strength they possess a hard endosperm. This is mainly due to the hot and dry weather conditions prevailing during the wheat maturing and harvesting period. Upon milling, the flour from these wheats has a very high damaged starch content leading to higher water absorption values for doughs. This quality characteristic is gainfully utilized for the preparation of good quality *chapaties* and other flat breads but may lead to serious problems in producing acceptable quality bread.

Wheat Usage in India

Wheat is now the second most important cereal grain crop of India and is cultivated on nearly 23 million hectares of land with an annual production of 54 million tons. Most of the wheat is grown in the northwestern part of the country. The important wheat growing states in India are: Uttar Pradesh, Punjab, Haryana, Madhya Pradesh and, Rajasthan. The surplus wheat from these states is procured by the state and central government agencies at a support price fixed by the

Agricultural Costs and Price Commission. The role of the public sector in handling food grains, primarily wheat, has increased both in terms of acquisition of stocks in response to the needs of the society, as well as to protect the interests of the growers. The volume of buffer stocks of food grains including wheat is maintained through imports and domestic procurement. Out of the 11 million tons of buffer stock of wheat maintained by the government of India, the state of Punjab alone contributes 65% of the total.

The wheat grown in India is mainly winter wheat. It is sown in October or November and harvested during mid-April to mid-May. The Aestivum wheats contribute a little over 90% of the total production from the area under wheat cultivation, whereas, durum (8%) and dicoccum (1%) wheats account for the balance (Hanslas and Sikri, 1989). The wheat cultivars grown in Northern India are high yielding with an average yield of 2 tons per hectare. The substantial increase in wheat production in India has been attributed to extensive irrigation facilities by way of a network of canals, tubewells, and pumping sets as well as adoption of high yielding cultivars responsive to high doses of fertilization. In the state of Punjab, the average yield of 3.5 tons per hectare for bread wheats is the highest in the country (Bakshi and Nanda, 1990).

Every year the 600 to 700 new strains developed under the All India Co-ordinated Wheat Improvement Project of the Indian Council of Agricultural Research, New Delhi are tested in various locations all over the country. The purpose is to identify superior cultivars for release to the farmers for cultivation on their fields. The hard working farmers of India deserve the credit for bringing about the "Green Revolution," raising the total wheat output from about 12 million tons in 1964-65 to over 54 million tons in 1991-92. The history of wheat breeding research and development has recently been summarized by Swaminathan (1990).

The Aestivum wheats are milled to whole wheat flour for use in the preparation traditional hearth baked products like *chapaties, poories, naan, and paronthas* with the stone mill (chakkies),or to low extraction flour using roller flour mills for the production of Western type products like pan bread, biscuits, cookies, crackers, cakes, etc. The durum and dicoccum wheats are milled into semolina for use in traditional products like *rawa idli, upma* and, *halwa*. A portion of aestivum and dicoccum wheats also finds use in the malted food industry.

Flat Breads

Wheat in India is consumed mainly in the form of unleavened flat bread known as *chapati*. Other indigenous products like *parontha, naan, kulcha, bhathura* and *poories* are not uncommon. India produces about 54 million tons of wheat annually. Roughly 75 to 80 percent of India's wheat is consumed in the form of such products. The per capita wheat consumption in India is 58 kg per year. *Chapati* is the staple food of a large segment of the population of northwestern India. *Chapaties* constitute an important source of dietary proteins, calories, some of the vitamins and minerals for the poorer section of the Indian population (Bajaj et al 1990).

Traditionally, wheat is milled into whole wheat flour *(atta)*, on a power driven stone mill *(chakki)*. After removing three to five percent of coarse bran by sieving, the 95–97% extraction *atta* is used for the preparation of various whole wheat flour products, as described below :

Chapati is the most popular unleavened flat bread in India and is consumed during almost every meal of the day. The two

Figure 8-1. Traditional method of *chapati* making, (a) dough is covered with a cotton cloth, (b) rounded dough ball lying in the dusting flour, (c) *chapati* being sheeted by rolling pin, (d) *chapati* being baked on hot iron griddle, (e) chapati being puffed on hot coals in the hearth, (f) prepared *chapaties* are wrapped in cotton cloth till served.

most important quality parameters for *chapati* are softness and flexibility. Dough is prepared (Fig. 8-1) by hand mixing of flour with an optimum amount of water. The dough is rested for 15–20 minutes at room temperature, depending upon the convenience of the preparer. About 60–100 g of the dough is rounded between the palms and sheeted manually into a disk 2–3 mm thick using a wooden rolling pin. It is then immediately baked on a preheated iron griddle (*tawa*) at a temperature of 230°C for 1 minute on each side (Sidhu et al 1988). Figure 8-2 gives the flowsheet for *chapati* making. As *chapaties* are quite thin, they are very susceptible to moisture loss as well as staling

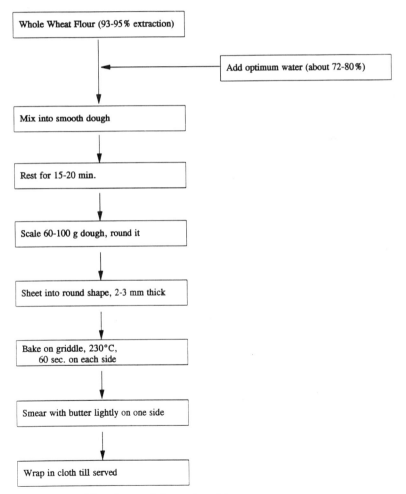

Figure 8-2. Flowsheet of *chapati* making.

after baking. Therefore, they are prepared fresh for both lunch and dinner. As consumers prefer to eat *chapaties* fresh and hot from the baking pan, *chapaties* older than 8–10 hours are not consumed by humans, but are fed to animals.

Chapati is also known by other names, such as *phulka* and *roti*. These are mere variations of the *chapati* in terms of product thickness and size. *Phulka* is about 12–15 cm in diameter and 1–2 mm in thickness, made from 30–50 g of dough. Roti is normally 2–3 mm thick and 20–25 cm in diameter, made from 130–160 g of dough. The other types of flat breads are: *roomali roti* and *tandoori roti*. *Roomali roti* is very thin (1 mm) and 30–35 cm in diameter and can be prepared only by an experienced person. The dough is prepared using patent wheat flour and water. About 50–100 g of dough is rounded, sheeted to 15 cm diameter and 2 mm thickness. The sheeted dough piece is extended to the ultimate desired size by flipping over the hands, and then baked over a preheated (230°C) convex iron griddle for 30 seconds on each side. It is prepared fresh and consumed as soon as possible. *Tandoori roti* is prepared from whole wheat flour/water dough. The *tandoori roti* is 12–15 cm in diameter and 3–5 mm thick. About 100–150 g of the rounded dough is sheeted on wet hands by an experienced person and baked by applying it to the inner side of a preheated *tandoor* (wood fired cylindrical hearth, made of clay, having a side hole near the bottom for air-inlet).

The very first impression of acceptability a consumer receives from *chapaties* is based on their color. A creamy yellow is the most desirable color for *chapaties* (Sidhu et al 1988). *Chapati* color is affected by the wheat cultivar, flour extraction rate and the processing treatments the flour has undergone. In urban Punjab, there is an increasing trend to the consumption of whiter *chapaties* made from wheat flour of lower (80 to 86%) extraction. Although the color of the *chapaties* is improved considerably by extensive debranning before milling into *atta*, the textural characteristics are affected adversely (Bajaj et al 1990). The color of *chapaties* is also affected by certain fungal diseases such as Karnal bunt. This disease, caused by *Neovossia indica*, a serious menace in the main wheat belt of India, affects the color of *chapaties* adversely. Washing fully infected grains improves the acceptability of subsequently produced *chapaties*, up to a level of 5% incorporation into sound wheat (Pal et al 1990).

The preparation of *chapaties* from blends of wheat and other cereals (corn, sorghum, millets) or pulses (chick pea) is quite common in different parts of the country. The effect of using mixtures of wheat and non-wheat flours for *chapati* making has been studied by Murthy and Austin (1963). Substitution levels can seldom exceed 25% without affecting color, texture and/or flavor adversely.

Chapati, being an unleavened, thin baked product, is highly susceptible to staling during storage. *Chapaties* become stale and quite unacceptable within 8 to 10 hours after preparation (Sidhu et al 1990). Although at present, *chapaties* are prepared fresh for the main meals, considerable demand for factory baked flat breads is steadily developing due to the urbanization of the country. There is a good possibility that commercial production of *chapaties* in India may start in the very near future. Under such circumstances, the product must reach the

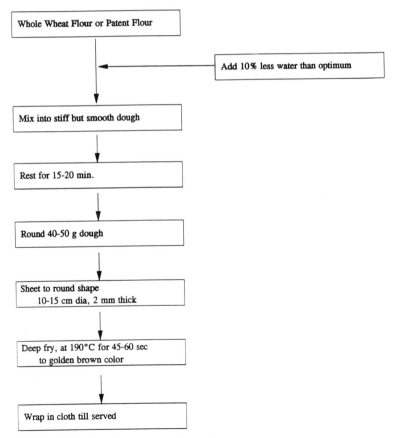

Figure 8-3. Flowsheet of *poorie* making.

consumers in an acceptable form (i.e. possessing an acceptable degree of freshness).

The major problem in the storage of all these products is textural deterioration. In spite of numerous efforts (e.g. inclusion of fat and/or surfactants), no satisfactory solution has yet been found to arrest their staling.

Poories are prepared using either patent or whole wheat flour with salt, fat and water. The dough is normally stiff, hav-

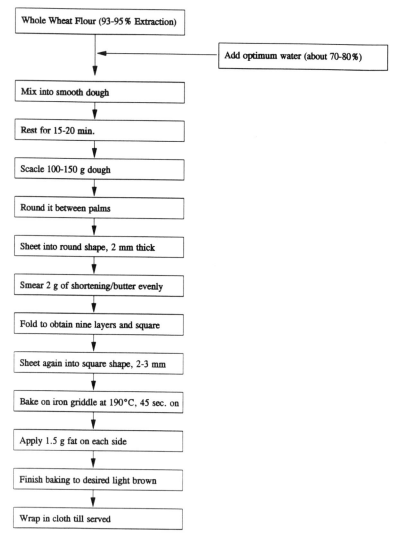

Figure 8-4. Flowsheet of *parontha* making.

ing about 10% less water than the optimum absorption. About 40–50 g of the dough is rounded, sheeted to 2 mm thickness and 10–15 cm diameter. It is then deep fat fried at a temperature of 190°C for 45–60 sec to a golden brown color (Sidhu et al 1990). Figure 8-3 gives the flowsheet for *poorie* making.

Parontha is a variation of *chapati* which has fat smeared between dough folds before sheeting and baking. The dough is prepared from whole wheat flour and water in a manner similar to that of *chapati*. Between 100 and 150 g of this dough is sheeted to 2 mm thickness and 15 cm in diameter. A total of three to five grams of fat is smeared on this sheeted dough, the sheet folded twice, and after applying the remaining fat, folded twice again to produce a total of nine layers. The folded dough is sheeted further (3 mm thick) to the desired basically square shape. Most *parontha* is square in shape but may also be round or triangular. It is then baked on a preheated iron griddle (190°C) for 45 seconds on each side. About 1.5 g fat is then applied on each side and the baking is completed to a desirable light brown color. *Parontha* can be prepared plain or stuffed. Stuffing is placed on the sheeted dough before folding. A number of variations can be created by using stuffings such as mashed potatoes, grated cauliflower, radish, chopped fenugreek leaves, minced meat, whipped eggs, acid coagulated green cheese, and cooked pulses. Figure 8-4 gives the flowsheet for *parontha*-making.

Naan. It is believed that the Muslim rulers who conquered India brought a new wheat product, *naan*, to the country. This type of product is also a very common form of bread throughout the Middle-East. *Naan* is a leavened product, traditionally prepared using low extraction flour, salt, water, buttermilk/curds/yoghurt/baking soda and/or baker's yeast, or eggs, boiled potatoes, or wild yeast from the atmosphere. The dough is mixed, fermented for 4–6 hours, then punched down if required. Fermented and punched dough is divided into 80–100 g pieces, molded and allowed to rest for 20 minutes. It is flattened to 3 mm thickness and baked like *tandoori roti*.

Mechanization of Indian Flat Bread Production

Organized commercial production and marketing of *chapati*, *poorie*, and *parontha* has not developed, mainly due to lack of appropriate indigenously built machinery and the high susceptibility of the products to microbial spoilage and staling. Mechanized production of *chapaties* involves five main unit operations: dough mixing, dividing, rounding, sheeting and baking. The machinery required for the first three unit operations is available indigenously. However, equipment is not available for proper sheeting and baking operations. Recently a number of German companies have supplied complete plants for the production of Arabic flat breads to the Middle-Eastern countries. After minor modifications, most of this machinery (except the baking oven) can be used for the commercial production of *chapaties*. As regards the baking operation, no satisfactory *chapati* baking oven is yet available.

Western Style Baked Products

Although less than 5 million tons of white flour is used for the production of western style baked goods, it represents one of the single largest organized processed food industries of India. Approximately 53,900 bakery units operate in the country with a total annual turnover of 7 billion rupees. The leading producers of bakery goods are registered under the Director General Technical Development (DGTD). Roughly 35 are registered for biscuit and 19 for bread production. The size distribution of these units is: large, 22; medium, 383; small, 3,728; and household, 49,767. Because of the growing population and increasing urbanization, there is great potential for growth in the Indian baking industry in the coming years (Radhakrishnan, 1987). Currently, the per capita consumption of bakery products is only 0.75 kg. in India as compared to 50 kg. in western countries and 150 kg. in some East European countries. However the total production of bakery goods by the large bakeries is expected to reach 3 million tons by the mid-1990s, out of which bread will constitute 50% and the balance include biscuits and other bakery items. The biscuit industry is the fastest growing segment of the market with an average rate of growth of 5 to 10% per annum (Datta, 1990). The various western style bakery goods produced commercially in India are discussed below:

White Pan Bread. There are over 600 roller flour mills in India with a total milling capacity of 15 million metric tons. However, the utilization of capacity has not reached even 50% of the total installed capacity, mainly because of the poor market demand for white flour for the production of western style baked products. Approximately 7 million tons of wheat are roll milled to produce the 5 million tons of flour which finds use in the production of pan bread, cookies, biscuits, crackers, cakes, pies, pastries and other traditional sweet and savory products. As per the 1983–84 estimate, the 1.03 million tons of pan bread are produced in India annually, and utilize only 75% of the installed capacity (Singhal, 1987). Due to the deregulation of the milling and baking industry by the Govt. of India during 1986, there has been expansion in the bread industry. A large number of bakeries in the organized sector are now mechanized. As a result, the production of bakery goods in India is projected to dramatically increase in the 1990s (Anon, 1989).

The Indian baking industry experiences difficulties in procuring the correct quality of flour for bread making (Datta, 1990). This problem arises out of the fact that the wheat procurement agencies do not segregate and store wheat cultivars according to bread making qualities. There is no system available to analyze the wheat arriving in the market; as a consequence, all the lots are bagged and stored at random. A definite and scientific grading system for wheat procurement and storage would be most useful in helping to supply the required quality of flour to the baking industry (Bakshi and Nanda, 1990). With such varying quality, bakers struggle to produce the uniform quality product demanded by the consumers. Moreover, in India all purpose flour is the standard produced by the millers. Lately, some of the mills have started marketing bread and biscuit flours separately (Radhakrishnan, 1987). Indian bakers normally use GMS as an emulsifier, calcium/sodium propionate as mold- and acetic acid as rope-inhibitors. Because sugar and shortenings are becoming expensive, there is a tendency on the part of bakers to minimize the use of these ingredients. Most bread manufacturers use flour, salt, yeast and water, plus ~1% sugar and shortening, and the required amount of mold/rope inhibitors for bread making.

Variety Breads. In India, the majority of the bread produced is white pan, sliced toast bread. However, a few bakers do produce small amounts of brown bread, fruit bread, high protein bread etc. Because of the constraints of machinery, technology and promotional efforts, the development of variety breads has been very slow in India.

Biscuits, Cookies, and Crackers. The size of the biscuit market in India is estimated to be nearly 0.5 to 0.55 million tons per year and growing (Datta, 1990). This particular segment of the bakery industry is growing at an average rate of 5 to 10% per annum. Bakery products, especially biscuits, cookies and crackers are best suited to satisfy mounting consumer demands without sacrificing profitability. While in the western countries, more than a thousand types of cookies, crackers and biscuits are produced, in India hardly a dozen types of these products are available. Among the biscuits being produced on a large scale the most common is the glucose type biscuit. This type of biscuit is highly subsidized by the industry in order to make it reach far and wide so that the brand names become familiar. Other types of products under this category are sandwich cream biscuits of different flavors. Sandwich cream wafers are also being produced by some bakeries. Among the cookies, the chocolate and coconut types are the most commonly produced. In the tea biscuit category, marie and krackjack are most popular.

Fermented crackers such as saltines and puffs are highly popular among consumers. Apart from these common categories, small bakeries produce a number of sweet and savory biscuits, by making use of common spices, candied fruits, nuts, oilseed meals and pulse flours.

In the western countries, the biscuit, the cookie or cracker manufacturer has the luxury of using a specific type of flour to suit each specific need. This is not the case for the Indian baker who has a very limited choice in the selection of flour and has to produce bakery goods by devising methods and means to overcome these shortcomings. Occasionally bakers have to make use of tuber/cereal starches or rice flour to adjust the strength of wheat flour.

Cakes and Sweet Goods. The Indian baking industry has ventured somewhat into large scale cake and sweet dough production. Lack of larger entry into the market is mainly because of the high cost of some of the required ingredients. A couple of commercial bakeries, however, have lately started marketing packaged, ready to serve, fruit cakes which are also produced by small scale bakeries for local sale in their areas of operation. The other types of sweet dough goods like pastries, donuts, cinnamon rolls, etc., are produced in very low volume by small bakeries for local consumption.

Although there is a great potential market for another class of convenience foods (i.e. ready bakery mixes), no such ready bakery mixes are yet available in India. A small beginning has recently been made in the development of bakery mixes for a complete cake mix, partially complete cake mix, chocolate cake mix and cake donut mix (Selvaraj and Sidhu 1986). The type of flour required for the production of these products is readily available in India so processors will not face any difficulty in procuring supplies. It is, however, the high cost of shortening, sugar and packaging materials which makes them expensive and, thus, hinders the growth of these type of products.

Breakfast Cereals. Though the list of extruded and other RTE breakfast cereals is extremely long in the western countries, it is almost non-existent in India. There exists, of course, a vast potential for the production of a variety of breakfast cereals based on wheat and other grains. Except for cracked wheat (*dalia*), rolled oats, corn and wheat flakes, no other breakfast cereal is available in the Indian market. Because of their relatively high cost, the use of these breakfast cereals is limited only to the affluent sections of the society. In common households, traditional Indian foods like *parontha, dalia, upma,* etc., are used as breakfast items. Recently, in the urban areas, white pan bread has also become popular as a breakfast item. At present, cracked wheat (*dalia*) is the important breakfast cereal in northern India. *Upma* (made from wheat semolina) is commonly served during breakfast in southern India.

Durum Wheat Products. Durum wheat cultivation is gaining popularity among Indian farmers, mainly because of its higher yield potential and resistance to common rusts. Durum production in India accounts for nearly 8% of the total wheat harvest. Moreover the macaroni products, particularly spaghetti, noodles and vermicilli, are becoming quite popular in the dietary pattern of the Indian populace. Accordingly, a large number of pasta processing units are being built to supply both local needs and for export.

Malted Food Products. Wheat malt extract is being utilized for the production of malted foods which are highly popular as ready-to-serve beverages for children (Bajaj and Sidhu, 1991). Wheat malt also finds use in the production of malted whiskeys as a means to partially replace more expensive barley malt. As the technology for the use of wheat malt and malt extract in the above foods is kept a closely guarded secret by the food processing units concerned, no scientific published information is available.

Traditional Sweet and Savory Products

There are a large variety of sweet and savory wheat based snacks across the length and breadth of the country. Products like *samosa, mathi, shakarpara, namkeenpara, gujiya, holige, modaka, kachori, sev, mattar*, etc., are popular snacks. Most are prepared at home or by small scale sweet/savory vendors. All these sweet and savory products are prepared by deep fat frying of a stiff wheat flour dough after shaping it into the desired form. Wheat flour is mixed with a small amount of shortening (3–5%), salt and sugar (2–3% each) and water sufficient to create a stiff dough. This dough is flattened to the desired thickness, cut into shapes and deep fat fried at about 130–170°C. The most desirable attribute of all these snacks for consumer acceptance, is their crisp texture. The amount of shortening added to the dough, moisture content of dough and frying temperature are the factors which determine the crispness of the snacks. Among the above snacks, *samosa* and *kachori* are the most popular savory products. Both have a stuffing, normally consisting of sliced potato, peas, pulses, spices etc. Because of the higher (70–80%) moisture content of stuffing, these products have a very poor shelf life. For these reasons, they must be prepared fresh and served hot, immediately after frying.

There has recently been a trend toward the development of ready mixes for some of traditional products (*upma, rawa idli, sooji halwa,* semolina or vermicilli *kheer, jalebi* etc). Production involves simple unit operations (e.g. grinding, roasting, blending, packaging). Low product moisture is mainly responsible for the mixes' reasonably good shelf life (six months to one year), though insect infestation and rancidity are major storage problems encountered in their marketing.

A large number of traditional sweetgoods like *jalebi, balushahi, sohan papadi, chiroti, pheni, ghevar,* etc., are also popular throughout the country. Patent wheat flour is one of the important ingredients in the preparation of all these. In some of the cases (e.g., *Balu shahi*), wheat flour dough is shaped, deep fat fried and dipped in sugar syrup. In other cases (e.g. *jalebi*), wheat flour batter is fermented, shaped and deep fat fried, and subsequently immersed in 50 to 60° brix sugar syrup. Most of these sweets are of intermediate moisture con-

Table 8-1. Composition of Some Wheat-Based Traditional Products

Product	Moisture (%)	Protein (%)	Fat (%)	Total Sugar (%)
Jalebi	9.9	2.57	12.9	49.7
Balu shahi	15.0	3.68	26.2	33.2
White *sohan papdi*	2.6	2.55	23.4	46.3
Samosha	49.8	9.43	13.6	...
Trisnack (salted)	3.5	7.80	40.5	...
Trisnack (sweet)	3.2	6.30	27.1	28.2
Halwa mix	1.5	6.00	31.8	37.0
Kheer mix				
Vermicilli based	3.3	14.00	4.5	49.0
Semolina based	1.7	14.40	4.5	50.6

Source: S. S. Arya, Indian Miller, 21(3):17-22, 1990.

tent and are susceptible to fungal spoilage. *Gulab-jamun* is a traditional milk-based Indian sweet. The use of wheat semolina/flour in the recipe (5/15 parts, respectively) has been suggested for obtaining improved desirable texture of the finished product (Minhas et al 1985). Data presented in Table 8-1 gives the composition of a few traditional sweet and savory products.

Recipes for a few representative sweet and savory snacks follow.

Balu shahi. Preparation of dough for *balu shahi* is quite similar to that of pie crust. However, it is not baked, but deep fat fried.

Ingredients:

Flour	500 g	Melted butter	220 g
Yoghurt	180 g	Sod. bicarbonate	2 g
Warm water	15 ml	Khoa	60 g
Rose essence	5-6 drops	For syrup:	
Shredded pistachio		Sugar (600 g)	
nuts	30 g	and water (325 ml)	
Shredded almonds	30 g	Oil/shortening for frying	

Sieve flour and soda together. Add melted butter, yoghurt and water and knead into a soft dough. Divide the dough into 35 equal parts and shape into round balls. Flatten the balls slightly and make a depression in the center. Deep fat fry in oil (shortening) at 150°C until light brown in color. Prepare the sugar syrup and cook until quite sticky. Arrange the fried, cooled product in a tray and pour the syrup over. Decorate immediately with shredded almonds, pistachio and *Khoa*. *Khoa* is whole milk, concentrated in an open pan to have 60% total solids content.

Mathi

Ingredients:

Flour	140 g
Shortening	35 g
Water	30 ml
Salt	to taste

Rub oil/shortening into flour thoroughly. Add water and knead into a hard dough. Rest the dough for 10–15 min. Sheet 50 g of dough to 6–7 mm thickness using a wooden rolling pin.

Dock with a fork so that it does not puff during frying. Fry in oil at 150°C to light brown color.

For *mattar* (another savory snack), the *mathi* dough is kneaded to a slightly softer consistency. After sheeting to 3 mm thickness, it is cut into 1 cm × 4 cm strips and deep fat fried to a light brown color and crisp texture. These are served as snacks during tea.

Gulab-jamun

Ingredients:

Skim milk powder	500 g	For sugar syrup:	
Flour	125 g	Sugar	800 g
Baking powder	3.5 g	Water	650 ml
Whipping cream	600 ml	Rose essence	2-3 drops
Oil for frying			

Make syrup with sugar and water. Add a spoonful of milk to the boiling syrup for clarification and remove scum which comes to the top. Cool the syrup after straining through cheese cloth.

Knead skim milk powder, flour, baking powder and whipping cream into a smooth dough. Rest the dough for 10–15 min. Divide into 15 g pieces and round between the palms. Fry these balls in oil at medium temp. (130°C) to an even brown color. Drain excess oil and immediately soak the fried balls in cold syrup overnight. Add rose essence to the syrup before serving. (One or two kernels of pistachio or raisins can also be rolled in the center of balls before frying).

Halwa suji

Ingredients:

Suji (semolina)	200 g	Unsalted butter	240 g
Sugar	200 g	Water	600 ml
Kesar (saffron)	2 g	Small cardamom	15 pieces
Chopped cashew nuts	50 g	Shredded blanched	
Raisins	50 g	almonds	50 g

Prepare a syrup by combining sugar and cardamom skins with water and stirring to dissolve. After the sugar is dissolved, boil rapidly for 5 min. Remove any scum by straining through cheese cloth. Set aside. After grinding the saffron, mix it into the strained syrup. Melt the butter, add semolina and roast it over a low temperature till the mixture becomes light golden in color and the melted butter starts to separate from the mass. Add the syrup and stir vigorously till the semolina absorbs all the water. Mix in the crushed cardamom seeds, raisins, shredded almonds and cashew nuts. Serve while still hot (Singh, 1975).

Khatai

Ingredients:

Flour or semolina	200 g	Powdered sugar	200 g
Unsalted melted butter	180 g	Yoghurt	70 g
Amm. bicarbonate	4.5 g	Almond essence	few drops
Grated almonds	25 g	Baking powder	5 g
Grated pistachio	25 g	Raisins or candied	
Chopped cashew nuts	25 g	fruit	15 g

Beat butter and sugar together until it becomes fluffy and creamy. Dissolve ammonium bicarbonate in a small amount of water and add to the creamed mass along with salt and curd. Continue beating until quite fluffy. Add the sieved flour or semolina, baking powder, almond essence, raisins or candied fruit and half of the grated almonds and pistachio. Shape the dough into a long roll (3 cm dia) and cut into one cm thick pieces. Coat the surface with the remaining chopped nuts and bake at 325°F for 15–20 min. The baking is complete when well risen, lightly colored and set. Remove from oven, cool and store in airtight containers.

Samosa is a deep fat fried product. It is one of the most popular snacks at tea time.

Ingredients for outer shell:

Flour	240 g	Water	140 ml
Shortening	60 g	Salt	1 g

Stuffing ingredients:

Potatoes	240 g	Paprika	3 g
Green peas	180 g	Dried pomegranate	
Ginger	35 g	seeds	3 g
Coriander seeds	4 g	Onions	70 g
Fresh coriander		Dried green mango	
leaves	15 g	powder	5 g
Cumin seeds	5 g	Shortening for frying	
Green chilies	10 g		

Stuffing: Boil the potatoes, peel and cut into half inch cubes. Fry chopped onions. Add green coriander leaves, green chillies, ginger, peas, salt, cumin seeds and paprika. Cook a few more minutes till peas become soft.

Shell: Sieve flour and salt. Mix shortening. Add water and knead into a smooth hard dough. Cover dough with a wet cloth and rest for 15 min. Divide dough into 40 g pieces and sheet into round shape, 2 mm thick. Cut each into two semi-circles. After filling with spoonful of stuffing, shape into a tetrahedron shape and seal the edges. Fry in oil at 180°C to a golden color. Serve hot with mint coriander chutney or tomato sauce. As a variation, cooked minced meat can also be used as stuffing.

Jalebi. This is a very popular sweet prepared from wheat flour. It is often served during marriage feasts.

Ingredients:

Batter:		**For syrup:**	
Flour	500 g	Sugar	800 g
Water	550 ml	Water	600 ml
Active dry yeast	5 g		
or compressed yeast	15 g	Oil/shortening for frying	

Dissolve sugar in water over medium heat and bring to a boil. Filter through cheese cloth to remove scum and cool the syrup.

Dissolve yeast in lukewarm water and add to flour. Add the remaining water and mix to a smooth batter. Allow the batter to ferment overnight. Beat the batter again to make it more homogeneous. Making a small pouch of tough cotton cloth (with a small hole in the center), the batter is extruded through the orifice directly into hot oil (140°C) in concurrent circular rings. Deep fat fry to a light golden color, drain excess oil and immediately immerse in syrup for 1–2 min, remove, drain excess syrup and serve hot.

Gol gappas are a very popular savory snack. They are a fried, puffed, crisp round ball shaped (3–4 cm dia) made from wheat flour/semolina or a combination of the two. After making a small hole on one side of the puffed product, cooked diced potatoes/chick pea/chopped onions are stuffed into it. Water spiced with powdered cumin, tamarind extract, dried mango powder and red chili powder is filled into the cavity and the product is eaten immediately.

Ingredients:

Semolina	100 g	Flour for dressing	50 g
Flour	100 g	Water	100 ml
or semolina alone	200 g	Oil/shortening	
Black bean flour	10 g	for frying	

Flour, semolina and black gram flour are kneaded into a hard dough. After a floor time of 15–20 min, dough is kneaded again. Taking about 100 g at a time, sheet as thin as possible with a rolling pin, using dressing flour occasionally. Cut into discs with cookie cutter. To avoid skin formation cover the discs with a wet cloth. Deep fat fry in oil at 150°C to light golden color. Drain excess oil on a sieve. These are served cold with spiced water.

Wheat Usage in Pakistan

Wheat, the staple cereal in the Pakistani diet, is grown all over the country, but more than 72% of the crop is harvested from the province of Punjab. In spite of good harvests over the years, the country still imports about two million tons of wheat annually. This situation has arisen because of the fact that wheat has now become the mainstay of animal feeds due to higher costs of all other conventional rations (barley, sorghum, *bajra,* maize, etc).

Wheat and its products are important components in the diet of the Pakistani population. The average per capita consumption of wheat in Pakistan, 164 kg a year, constitutes about 83% of the total cereal intake and provides over 50% of the total calories and 60% of the total proteins. Wheat alone can meet the daily needs of thiamine and niacin and more than half of daily requirements of iron and riboflavin. Cereals are, therefore, considered to be the main and the cheapest source of proteins and calories for the Pakistani population. About 80% of the total wheat is consumed in the form of *chapaties*, while the remaining goes for other bakery products. Annual wheat production in Pakistan is slightly over 12 million tons. Of this total, about 70% is used for human consumption and seed purposes (Chaudhry et al 1987). Pakistani and, for that matter, Punjab wheats are known as Karachi wheats in the national and international wheat trade. These wheats are noted for high flour yields, high water absorption and good dough characteristics.

As in India, wheat is used mainly for milling into *atta* for *chapati* production. Therefore, an amber grain color is strongly preferred. Because of this, the majority of the varieties are bred for amber color and plump grain quality. *Chapaties* which have creamish white color, soft silky texture and sweet wheat-ish aroma, are preferred by the consumers. The country's breeding programs are accordingly oriented to select wheat varieties conforming to these traits.

Chapati. In Pakistan, 80% of the total wheat produced is consumed in the form of unleavened flat bread (*chapati*). The wheat grain is ground into whole-meal (*atta*) including bran and germ for the production of *chapaties*. The *chapati* making procedure is similar to that followed in India. Although the *chapati* is a major food in Pakistan, no uniform and standardized method of preparation of *chapaties* exists and the technique varies from place to place and household to household.

Although *chapaties* are usually made from whole wheat flour, the use of other cereals, pulses and tubers to enhance their nutritional and sensory quality is not uncommon. Sprouted triticale can enhance the protein contents of "*dalia*," "*halwa*," "*Sohn halwa*," "soybean-triticale milk," "curd" and "lemon-wheat squash." The blending of triticale with Chenab-70 wheat cultivar at 25–50% levels resulted in highly acceptable baked products. In case of *chapati*, the product is richer in protein and lysine than those made from commercial wheat alone. As an added advantage, *chapaties* made from triticale alone or wheat-triticale blends are slightly more resistant to staling. Ullah et al (1979) have documented the use of cereals other than the traditionally used wheat for the preparation of *chapaties*. Blends of wheat with rice, barley, sorghum, maize and millet can be used for *chapati* making. *Chapaties* prepared from wholemeals of sorghum, maize and pearl millet, when used alone, are of poor quality but improved (in the descending order) when these cereals were blended with coarse rice (1:1 ratio).

In spite of the massive amounts of wheat utilized in this single product of daily use, no serious attempt has been made to standardize a uniform test baking procedure for *chapati*. Even the single factory producing *roti* on a commercial scale has not succeeded as a viable alternative to household baking of *chapaties*. Probably the consumers of *chapaties* in Pakistan have not yet abandoned the age old concept of eating this product fresh from the hot plate (baking griddle). Little scientific information is available about the preparation of other wheat based products, like, pan bread, biscuits, cookies, cakes, pasta, extruded cereals etc., though one to two million tons of wheat find use in these types of products. More research is needed to be carried out on the popularization of commercially produced *chapati* and other baked products in Pakistan.

Wheat Usage in Bangladesh, Sri Lanka, and Nepal

Bangladesh, which lies between 21°34' and 26°38'N latitude and 88°45' to 92°40'E longitude is roughly 80% flood plain, which is served by three major rivers. Wheat is the second most important cereal crop in this predominantly rice growing country. Wheat growth area and production have increased substantially in the country since 1974-75. During 1974-75, the area under wheat cultivation was 125,000 hectares which rose to 591,200 hectares within the short span of 5 years. During the same period, the production of wheat increased nine fold and has now more or less leveled off at about one million tons per annum. As all the available arable land is now cropped, greater emphasis is being laid on breeding and crop management for increased yields.

Bangladesh imports about 1.5 to 2.0 million tons of wheat annually depending upon the quantity of locally produced foodgrains. About 75% of the wheat is consumed in the form of *chapati, roti, naan, parontha* and *poorie.* A number of wheat based snack foods and confectionery items (quite similar to those available in India and Pakistan), *sooji halwa, jalebi, samosa,* etc., are also popular. About 20 to 25% of the wheat is milled into white flour for the production of western style products like bread, biscuits, cookies, crackers and cakes (personal communication, Ekramul Haque).

Sri Lanka is a rice producing and consuming nation. The per capita wheat consumption in Sri Lanka is only 44 kg per year. As the country does not grow any wheat, all the requirements are met by imports. Because of its lower price, primarily soft wheat is imported from France, Canada, Argentina, United States, Australia and Saudi Arabia (Anon. 1991). Sri Lanka normally imports 766,000 tons of wheat annually. It is milled into only one type of all purpose white flour (with 10.5% protein), 75% of which is used for producing western style products like bread and 25% for biscuits, crackers and noodles etc. (Anon, 1993).

Nepal grows about 750,000 tons of wheat annually and imports the required quantities to meet the demand. Most of the wheat is consumed in the form of *chapati* and other traditional products similar to India. A small amount of wheat is also milled into white flour to produce western style products like bread, biscuits, crackers and cakes.

References

Anon. 1989. Food Raw Materials for processing sector and future scenario in 2000 A.D. Indian Miller 20(3): 9-19.

Anon. 1991. World Grain Statistics. International Wheat Council, London, England, pp 2-4.

Anon. 1993. Strategic Plan FY-1993. US Wheat Associates Inc. Washington, D.C. pp D3-D5.

Bajaj, M., Kaur, A., and Sidhu, J. S. 1990. Studies on debranning of Punjab wheats. II. Effect on rheological characteristics and chapati making quality. J. Plant Sci. Res. 6(1-4):40-42.

Bajaj, M., and Sidhu, J. S. 1991. Development of wheat malt extract based carbonated beverage. Proceedings, Research Evaluation Committee, Punjab Agril. University, Ludhiana, India.

Bakshi, A.K., and Nanda, G.S. 1990. Wheat-quality improvement (Industrial and Nutritional) programs at PAU to match with changing requirements of millers and processors. Indian Baker 21(2): 21-30.

Chaudhry, N. M., Ullah, M., and Anjum, F. M. 1987. Effect of storage conditions on grain quality characteristics of wheat. Pakis. J. Agril. Res. 8(1): 17-23.

Datta, J. M. 1990. Quality flour requirements for bakery products. Indian Miller 20(4): 50-56.

Hanslas, V. K., and Sikri, S. 1989. Choose superior wheat varieties for specific end-uses. Indian Baker 20(1): 11-19.

Minhas, K. S., Rangi, A. S., and Sidhu, J. S. 1985. Indigenous milk products. II. Effect of recipe on the chemical composition of gulab-jamun. J. Fd. Sci. Technol. 22(4): 244-246.

Murthy, G.S., and Austin, A. 1963. Studies on the quality characters of Indian wheat with respect to chapati making. Food Sci. 12:61-64.

Pal, S., Saini, D. P., Malik, S. K., and Gautam, P. L. 1990. Effect of Karnal bunt on grain quality components and chapati making quality. Indian Miller. 21(6):23-26.

Radhakrishnan, K. L. 1987. Problems of baking industry with regard to raw materials. Indian Miller 18(3):15-27.

Selvaraj, A., and Sidhu, J. S. 1986. Development of ready bakery mixes. Proceedings, Product Development and Research Utilization Committee, CFTRI, Mysore, India.

Sidhu, J.S., Seibel, W., Bruemmer, J. M., and Zwingelberg, H. 1988. Effect of flour milling conditions on the quality of Indian unleavened flat bread (chapati). J. Fd. Sci. 53(5): 1563-1565.

Sidhu, J.S., Seibel, W., and Meyer, D. 1990. Gelatinization of starch during preparation of Indian unleavened flat breads. Die Staerke 42(9):336-341.

Singh, Balbir. 1975. as in "Indian Cookery", by Mrs. Balbir Singh, Mills and Boon Ltd., 17-19 Foley Street, London, W1A 1DR.

Singhal, S. C. 1987. Development in Bakery and confectionery fats. Indian Baker 18(1): 24-31.

Swaminathan, M.S. 1990. In: Wheat Research in India, 1966-1976. P.L. Jaiswal, ed. ICAR Publication, New Dehli, India.

Ullah, M., Bajwa, M. A., Ali, M. S., and Anjum, F. M. 1979. Superior quality chapati from blended cereals. Annual report on wheat research and production in Pakistan, Pakistan Agricultural Research Council, Islamabad, pp 235-245.

CHAPTER 9

Wheat Usage in the Middle East and North Africa

Jalal Qarooni

Introduction

The exact location of ancient wheat cultivation is not known with certainty. Archaeological excavations indicate that grain cultivation started either in Syria-Palestine or, a little farther to the north, along the southern parts of Anatolia. Drawings from the Fifth-Dynasty tombs of Ty and Mereruka, 2600–2500 B.C., show farm activities indicating cereal (emmer wheat and barley) growing and harvesting in ancient Egypt (Storck and Teague, 1952).

Excavation of the oldest baker's oven in the world showed that bread was known in Babylon in 4000 BC. In the old Kingdom of Egypt, bread was baked in hot ashes or on heated stone slabs and on the wall of barrel-shaped ovens (Pomeranz, 1987). Figure 9-1 shows a scene of pot (pan) breadmaking in ancient Egypt (Chazan and Lehner, 1990).

Both regions, the Middle East and North Africa, have diverse and variable climatic conditions. Cereals are grown in two major agricultural zones, the Mediterranean and the highland. The Mediterranean zone is hot and dry during the summer and cool and moist in winter. Cereals are grown here during the winter because of relatively high rainfall. The highland zone (1000 m above sea level) is characterized by cold winters and hot summers (ICARDA, 1983). Highlands cereal yields average about 650 kg/ha. Crops are generally planted in fall, although spring plantings are important in parts of Tur-

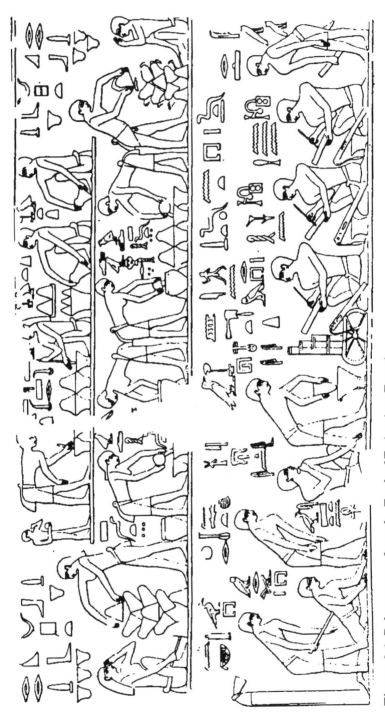

Figure 9-1. Bakery scene from the Tomb of Ty (ancient Egypt).

Table 9-1. Area of Wheat Cultivation, Yield, and Production of Middle East and North African Countries During 1987–1988 to 1989–1990[a]

Country	Area[b]			Yield/ha (MT)			Production (1,000 MT)		
	1987–1988	1988–1989	1989–1990	1987–1988	1988–1989	1989–1990	1987–1988	1988–1989	1989–1990
Afghanistan	2,000	2,100	2,000	1.38	1.33	1.35	2,750	2,800	2,700
Algeria	1,511	1,023	1,255	0.78	0.60	0.68	1,175	615	850
Bahrain	...	...	...	...	...	...	...	...	...
Djibouti	...	...	...	...	...	...	...	...	...
Egypt	577	597	630	4.23	4.76	5.05	2,443	2,839	3,183
Ethiopia	740	750	725	1.08	1.20	1.10	800	900	800
Iran	6,100	6,300	6,300	0.98	1.08	1.08	6,000	6,800	6,800
Iraq	859	950	550	0.84	1.05	0.89	722	1,000	490
Israel	90	93	90	3.31	2.27	2.23	298	211	201
Jordan	125	115	65	0.87	0.87	0.69	109	100	45
Kuwait	...	...	...	...	...	...	...	...	...
Lebanon	6	10	10	3.0	1.40	1.40	18	14	14
Libya	310	320	300	0.61	0.42	0.48	190	135	145
Morocco	2,288	2,317	2,630	1.06	1.73	1.49	2,427	4,019	3,927
Oman	...	...	...	...	...	...	...	...	...
Qatar	...	...	...	...	...	...	...	...	...
Saudi Arabia	630	666	645	3.81	4.80	4.81	2,400	3,200	3,100
Somalia	...	...	...	...	...	...	...	...	...
Sudan	150	200	250	0.93	1.25	1.20	140	250	300
Syria	1,183	1,100	744	1.40	1.88	1.21	1,656	2,067	900
Tunisia	971	299	557	1.4	0.74	0.75	1,360	220	420
Turkey	8,700	8,750	8,700	1.49	1.71	1.32	13,000	15,000	11,500
U.A. Emirates	...	...	...	...	...	...	...	...	...
Yemen	78	75	75	1.22	1.36	1.32	84	95	90
Total	16,318	25,665	25,526	1.67[c]	1.67[c]	1.59[c]	35,572	40,265	35,465

[a] Preliminary.
[b] 1,000 hectares, harvested area as far as possible.
[c] Average.

Source: Agricultural Statistics, 1990, United States Department of Agriculture, U.S. Government Printing Office, Washington, DC 20402

key and Afghanistan. Common wheat (*T. aestivum*) is the major crop in both regions. However, durum wheat (*T. durum*) occupies a large area, particularly where rainfall is limited to 300 to 500 mm (Briggle and Curtis, 1987).

About 21 million hectares of wheat are grown in Algeria, Egypt, Libya, Morocco, Tunisia, Angola, and Nigeria. Except for Egypt, the crop is rain-fed and yields average about 850 kg/ha. Because most of the wheat produced in Egypt is grown under irrigation, yields average more than 3.0 tons/ha. In North Africa, most of the cereal-growing area is planted to durum wheat (50%) and common wheat (20%) with barley occupying most of the remaining area. Wheat is grown as a winter crop, which coincides with the rainy season. It is planted in the fall and harvested in the May or June (Briggle and Curtis, 1987).

Today, wheat production in most countries of the Middle East and North Africa is below their consumption level and, hence, these countries (with a few exceptions) import substan-

Table 9-2. Major Importers and Exporters of Wheat and Flour (Wheat Equivalent) in the Middle East and North Africa, 100 MT

Country	Imports			Exports		
	1988	1989	1990	1988	1989	1990
Afghanistan	2,315	2,774	3,168	...	...	...
Algeria	38,570	58,996	38,343	...	...	...
Bahrain	409[a]	309[a]	308[a]	...	...	...
Djibouti	233[a]	197[a]	208[a]	...	...	...
Egypt	72,394[a]	69,707[a]	66,145[a]	...	9	...
Ethiopia	10,728[a]	4,195[a]	6,581[a]	...	...	...
Iran	28,341	45,000	44,200[a]	...	...	...
Iraq	29,389	34,389[a]	18,694[a]	...	...	...
Israel	6,032	5,590	6,805[a]	...	11[a]	10[b]
Jordan	4,291	1,791	6,243	120	233	376
Kuwait	1,628	1,990	1,014[a]	6	1	...
Lebanon	3,789[a]	4,083[a]	1,833[a]	...	...	...
Libya	6,745[a]	7,792[a]	10,695[a]	...	...	...
Morocco	14,689	12,900	14,136	24	...	...
Oman	952	1,315	1,404	13	15	183
Qatar	612	456	489	...	...	...
Saudi Arabia	1,601	1,398	2,681	16,113	16,113	15,069
Somalia	1,372	1,062	1,158[a]	...	...	...
Sudan	6,500[a]	5,195	5,404[a]	...	...	...
Syria	8,899	13,565	16,271	70	112	150
Tunisia	13,517	11,563	8,976	...	...	...
Turkey	100	20,368	21,876	25,386	8,316	1,759
U.A. Emirates	1,109	639	1,085	21	5	36
Yemen	11,099	12,548[a]	17,111[a]	...	...	...
Total	265,314	317,819	294,828	46,257	24,950	17,483

[a]Unofficial figure.
[b]FAO estimate.

Source: Food and Agricultural Organization (FAO), Trade Yearbook, vol. 44 (1990)

tial amounts of wheat. During the past 20 years, both production and consumption of wheat increased at an annual rate of 3.1% worldwide and 5% in developing countries (CIMMYT, 1987–88). However, if China and India are excluded, wheat consumption increased by 4.1% annually in the developing world against a growth in production of 3.3%, indicating a sharp drop in self-sufficiency in wheat in the Third World.

The area under wheat cultivation, yield per hectare of land, and production of the Middle East and North African countries are shown in Table 9-1. With the exception of Saudi Arabia and Turkey, wheat production in these countries does not meet the domestic consumption, hence, they import large amounts of wheat from exporting countries. Table 9-2 shows the major importers and exporters of wheat among these countries.

Wheat Production and Usage

Afghanistan. There are no recent data on the status of wheat production in Afghanistan. According to an FAO report (1976), wheat production in Afghanistan acounts for ~60% of all grain produced. Wheat is grown on almost every farm. About one fourth of the farmlands' crop is marketed, mostly in the cities, the remainder consumed by village inhabitants. Both winter and cold-tolerant spring wheats are grown in the valleys where they normally are covered by snow during winter. More than half of the area devoted to wheat cultivation is irrigated. Yields under irrigation are substantially greater than those from rain-fed fields (Yamazaki et al., 1981).

Iran. About 45% of the wheat growing area in Iran produces winter or semi-winter varieties, much of it in the highlands of Azarbaijan, Khorasan, and Kermanshahan. High yielding cultivars such as Bezostaya, a hard red winter selection introduced from the former USSR, have out-yielded local cultivars, such as Roshan and Omid (Hafiz, 1972). The yield of 12 Iranian wheat cultivars (Maleki, 1984) ranged from 1.22 to 6.28 tons/hectare. The highest yield was for the cultivar Kouhrang, with 41.4 g thousand kernel weight.

Israel. Wheat is intensively cultivated with modern methods. Average annual wheat production is 250,000 tons. Israel imports twice that amount mainly through the port of Haifa from exporting countries, especially the United States (World Grain 1991).

Jordan. Wheat production in Jordan covers only 16% of domestic consumption (Bahram, 1986). In a survey of locally grown cultivars, Amr (1988) reported that the cultivar Horani exhibited the highest protein content, and Asad 65 showed the highest milling yield. It was concluded that the local wheats performed poorly in respect to protein content and milling yield. The same author reported that only two cultivars, Asad 65 and Strok S, showed acceptable rheological properties for use in bread making. According to estimates by 185 durum wheat producers (Elings and Nachit, 1991), the average grain yield was lowest in the western mountainous regions and highest in the southern parts near Syria.

Saudi Arabia. For centuries, small farmers grew wheat with irrigation near the oasis of Qassim, in Saudi Arabia. The production ranged from 39,000 to 205,000 tons in the 1970s. In 1980, the government launched a program to become self-sufficient in wheat by offering $1,032 per ton to Saudi farmers and investors (Anonymous, 1990). Consequently, 350 circles of land (each covering some 52 ha) are now watered by center-pivot irrigators and yield from 6 to >8 tons/ha (Sjerven, 1989). Wheat production doubled by 1982 and nearly doubled again in 1983. By 1984, the country was self-sufficient in wheat. Wheat production in 1989–90 reached slightly over 3 million tons, a level greater than self sufficiency, and the government subsidy payment dropped to $309 per ton in 1990 (Anonymous, 1990). In addition to export to the members of the Gulf Cooperation Council (GCC), surplus Saudi wheat is exported to Indonesia, China, and European countries.

Khatchadourian et al. (1985), reported that the cultivars Yecora Rojo and Probred (red spring wheats) accounted for 95% of wheat production in the country. The rheological properties of flour from Yecora Rojo wheat was studied by Al-Mashhadi et al. (1989) who reported that milling yields of the cultivar grown under 12 different fertilization programs ranged from 65.4 and 67.8%, and the farinograph water absorption ranged between 68.8 and 71.2%.

Turkey. Wheat is cultivated in Turkey on the central Ana-tolean plateau and to the southeast and east, the Aegean and southern coast areas, and in Thrace. The central, southern, and eastern parts of Anatolia, the principal wheat producing area, have a continental climate and thus, are suitable for the cold-tolerant hard winter and durum wheats, the latter mostly in the southeast. Much of the crop is rain-fed, so production depends somewhat on sufficiency and timeliness of rainfall (Demirlicakmak, 1972). The Mediterranean climate along the coast areas, with abundant winter rainfall, has favored culture of spring wheats such as those of Mexican, French, and Italian origin. In order of commercial importance, the five principal types of wheat grown in Turkey (Seeborg, 1974) were: common white (soft to semi-soft); compactum; durum; hard red winter of the Bezostaya type; and the Mexican type of soft red wheat.

In recent years, the area under wheat cultivation in Turkey has reached approximately 9 million hectares, yielding an average 1.0 ton per hectare. Turkey exports wheat that is grown in the eastern regions to countries such as Syria, Iraq, and Iran. This export allows Turkey to relieve its storage space limitation by selling off soft wheat varieties at harvest time and replacing them with imported hard wheats (US Wheat, 1986–87).

North Africa. The potential land for cereal cultivation in 13 African countries including Algeria, Egypt, Ethiopia, Libya, Morocco, Sudan and Tunisia is about 53.9 million ha (Choudhri et al., 1987). Bakhella et al. (1990) studied the grain hardness of 18 common wheat cultivars grown in Morocco. They concluded that three cultivars, Sais (1615), Pinyte (2306), and Potam, could be classified as soft wheat, but all other cultivars were hard.

A projection of supply and demand for the Egyptian wheat production for the years 1990 and 1995 (Al-Hazek, 1988) indicated that it would be difficult achieving self-sufficiency even to the level of 40%. Similarly, the present wheat yield in Libya is 0.5 ton/ha (Malik et al., 1981). This is well below the average yield in Africa. Ethiopia is not self-sufficient in cereals (Mohammed, 1988), even though 4.7 million ha of land is under cereal production. Yields for most cereals are low and production is almost exclusively rain-fed.

Milling and Baking Industries

Most countries of the Middle East and North Africa import needed wheat and flour through the public (government) sector. This sector also handles the bulk milling (US Wheat, 1991–92). Prices of wheat and flour are subsidized by the governments, and the price of bread (the main staple) is controlled as well. This control, however, does not apply to pastries and other baked products.

The baking industry itself is predominantly private (US Wheat, 1991–92). However, governments regulate final product weight of the most popular breads.

Afghanistan. Large flour mills produce predominantly (97%) whole wheat flour (*atta*) and only 3% white flour. The production of smaller country mills is *atta*. The most popular breads are *naan* and *chapati* (Nagao, 1981).

Algeria. Wheat is the major cereal consumed in Algeria. There are presently about 114 flour mills (including durum) in the country (U.S. Wheat 1991–1992). Fourteen units of the existing mills were upgraded in 1992. These units have increased Algerian common and durum milling capacity by 1.7 million tons annually. Until recently, Algeria imported about 800,000 tons of semolina per year. The objective of that expansion was to reduce or eliminate semolina imports from Italy.

Algeria's baking industry consists of three major industrial plants (owned and operated by the government) plus hundreds of small bread and pastry shops. The biscuit (cookie) industry consists of 10 privately owned companies and six government operated plants. The government also operates 13 pasta and couscous plants.

Egypt. The flour milling industry consist of 200 roller and stone mills with a combined capacity of 15,200 tons/day. Sixty percent of the milling is done by roller mills and the rest by stone mills. Eight public sector milling companies own 90% of the country's capacity. Imported wheat is milled to flour of two extractions, 72% and 82%. These flours along with imported flour (72% extraction) are used for the production of local and European type breads, cookies, and pastries. The Grand Cairo Baking Company has a large number of automated, semi-automated, and manual bakeries producing the predominant local breads (*balady* and *shami*) (US Wheat, 1991–92). Private bakeries are involved in the production of a variety of local and foreign pastries.

Iran. Flours of specific extraction rate are milled for the production of local flat breads: 78% extraction is used for *barbari*, 82% for *lavash*, 84% for *taftoon* (*tanoor*), 87% for *sangak*, and whole meal (97%), for village bread (Faridi et al., 1981). The breads are produced three times a day in small bakeries, a common practice in countries where such single-layered flat breads are popular. Although large-scale, automated production lines exist, the freshly baked products from small bakeries are the most popular. Smaller amounts of flour of low extraction rates are used for the production of various types of cookies and pastries.

Iraq. Wheat and flour imports into Iraq are controlled by the State Organization for Grains and Foodstuff. Both traditional and modern types of milling and baking industries exist. The total daily capacity of the 13 public sector and 60 private mills is approximately 7,500 tons (World Grain, 1988). However, trained manpower is lacking to operate and manage the modern facilities (US Wheat, 1986–87). The most popular breads in Iraq are *khobz* (*tanoor*) and *samoon.* These are produced by hundreds of small bakeries. Imported, automated, flat-bread bakery lines also produce large amount of flat breads daily.

Israel. Israel has 22 flour mills with an annual total capacity of 1 million tons. According to the Israeli Milling Association, the annual flour production in recent years is 700,000 tons (World Grain 1991). The most popular flat bread baked in towns is *kimj.* White European bread prepared from flour of 75% extraction is produced in many varieties such as pan bread, Pullman loaves, and French and Vienna breads (Adler, 1958).

Jordan is one of the more technologically developed countries in the region, with a relatively high standard of living. This has been shown in the consumers' demand for a variety of breads and other baked products (US Wheat, 1991–92). The six flour mills of Jordan have a total daily capacity of 2,000 tons. In Irbid, the "National" flour mill has a daily capacity of 500 tons (World Grain, 1987). Milling technology is well developed and the current capacity exceeds the demand. The baking industry is privately owned and operated.

Morocco. Eighty-four industrial mills operate in the country. The current milling capacity is about 3.2 million tons annually, of which about 50% is located in the Casablanca, Fes, Meknes, and Kenitra areas. The country's milling capacity is planned to increase to nearly 4 million tons by the year 2000. Flour is sold at subsidized prices (US Wheat, 1991–92). The average flour extraction rate is 77%, but "deluxe flour" is produced with extraction rate of 68%.

The baking industry in Morocco is unique. There is a public oven for every 150–200 customers. People prepare their own dough and bake it at these public ovens. Consequently, commercial bakeries provide only 30% of the bread in the market, and their main customers are hotels and restaurants (World Grain, 1987).

Durum is the essential ingredient for the production of *couscous,* the traditional Moroccan dish. Various types of pasta, including spaghetti, are made either exclusively from soft wheat or blends with a small portion of durum. There are six biscuit (cookie) plants and 10 small shops producing about 20,000 tons annually. Most small shops make a variety of cakes and pastries. However, Moroccan consumers prefer quality home-made cookies for all ceremonial events (US Wheat, 1991–1992).

Saudi Arabia and the Gulf Countries. The Grand Silo and Flour Mills Organization of Saudi Arabia established the nation's first flour mill in Riyadh in 1977. By the end of 1986, this organization managed a wheat milling operation of 4,500 tons a day in various cities of Saudi Arabia. Flours of four extraction rates (70, 75, 85, and 95%) are produced in these mills (World Grain, 1985). The baking industry in Saudi Arabia produces a wide range of foreign and traditional products. In 1985, baked products were supplied by 100 fully automated, 500 semi-automated, and 4,250 manual bakeries. Over 90% of bread production is in large automated bakeries (Mousa and Al-Mohizea, 1987).

The Gulf States of Bahrain, Kuwait, Oman, Qatar, and the United Arab Emirates have highly developed milling, and baking industries. Although some variation exists among the grist and flour extraction rates, wheat is milled to low (75–78%, white flour) and high (90% and whole meal) extraction rates. High extraction flours are used for the production of local flat breads and many types of homemade baked goods and low extraction flour for the production of cakes, various pastries, and European breads. The price of flour is subsidized by the government in many of these countries. The traditional flat bread (*tanoor*) is produced in small bakeries, three times a day. The production and acceptability of Arabic bread (double-layered flat bread) has increased in the past decade, due to the availability of large automated production lines. A large variety of local and European style baked products, pasta, and extruded products are available in the supermarkets and shops.

Tunisia. The milling industry of Tunisia constitutes 21 flour and durum mills, which operate at 78% of their capacity. The baking industry consists of a large number (1,810) of small bakeries. There are 10 biscuit (cookie) plants in the country. The Tunisian pasta industry consists of many small pasta plants with 15 of industrial size. Three of these large plants supply 70% of the market's need (US Wheat, 1991–1992).

Turkey. In 1984, there were 367 mills in Turkey. Four of these mills had an annual milling capacity of 100,000 tons. Forty-one mills had an average annual capacity of 74,500 tons, and 330 mills had 30,000 ton capacity. During the harvest, ~3,500 stone mills are used by the farmers (World Grain, 1987). Baked products are reaching levels of sophistication that are close to European standards, but the variety of end-products is limited by lack of practical production knowledge or awareness on the part of the millers and bakers (US Wheat, 1991–92).

Yemen. The ministry of Supply and Home Trade, in conjunction with a committee made up of Yemeni government and private sector trading companies, have the total responsibility for all wheat and flour purchases for the country (US Wheat, 1986–87). The Red Sea Flour Mill, the country's main grain processor, with a total milling capacity of 1,900 tons daily, is the main supplier of locally produced flour to the cities of Hodaidah, Sanaa, and Taiz. Two types of flour are produced from wheat, white flour (72–75%) and flour of 88% extraction. Two large, automated Arabic and pan bread lines owned by the General Corporation for Foreign Trade & Grains are in operation in Sanaa and Taiz. In the cities, *roti* bread (half the size, in width and height of European pan bread), is a much demanded product. Although Arabic bread is produced in the local bakeries, *roti* is somewhat more popular. In the villages, *tanoor* bread (locally called *maloug*), a product which traditionally was prepared from barley flour, is produced from wheat flour and is the dominant product. Private bakery shops generally have some varieties of Middle Eastern cakes and pastries as well as European products.

Wheat-Based Products
in the Middle East and North Africa

Breads

Bread types of the world have been classified (Faridi, 1988) into three groups based on their specific volumes (volume / weight): (a) those with high specific volume such as Western pan bread, (b) those with medium specific volume such as French and rye breads, and (c) those with low specific volume such as flat breads. All three groups of breads are produced and consumed in the Middle East and North Africa. However, the popularity of pan bread is limited to the cities and major urban areas. The French-type breads and similar products are the most popular breads in many parts of North Africa, especially Algeria, Libya, Tunisia, and Morocco (Patel and Johnson, 1975; Finney et al., 1980). *Roti*, a French type bread baked in a small pan, is the most popular product in urban areas of Yemen. *Samoon* and hamburger buns (round and elongated shapes) are popular in the cities of Middle East and are utilized for sandwich making. The most popular breads of the Middle East and North African countries are listed in Table 9-3.

The production of flat breads in the Middle East and North Africa exceeds that of other types. The flour usage for flat bread production in metropolitan Kuwait, where the population is mainly from the Middle East and North African countries, is 85% of the total usage.

Flat breads in general, and those of the Middle East and North Africa in particular, can be divided into two groups according to their cross section: (a) single-layered flat bread such as Afghani *naan, bazlama, barbari, lavash,* Moroccan, *tanoor,* and Tunisian bread, and (b) double-layered flat bread such as Arabic (pita) and *baladi* breads. The thickness of flat bread varies from one type to another and within one type from one country to another. In general, their thicknesses range from paper thin (*rogag* and *sauge*) to 3–5 cm (*bazlama, barbari,* and Moroccan).

The main processing difference between the single- and double-layered flat bread is the final proofing period. In the double-layered flat bread, this period might exceed 30 min,

Table 9-3. Breads of the Middle East and North Africa

Country	Bread Type
Afghanistan	Naan, Chapati, and European bread
Algeria	Matlowa, French bread, Khobz El-daar, European bread
Bahrain	Tanoor, Arabic, Chapati, European bread, Samoon
Djibouti	Kisra
Egypt	Baladi, Shami, Samoon, French, Fatier, Shamsi, Bataw, European bread
Iran	Barbari, Tanoor (Taftoon), Lavash, Teeri, Suage, Sangak, European bread, French bread
Iraq	Tanoor (Khobz), Arabic, Samoon, European bread
Ethiopia	Injera
Israel	Sadj, Tarboon, Kimaj (Arabic), European bread
Jordan	Armani, Mafrood, Sauj, European breads
Kuwait	Tanoor, Arabic, European breads
Lebanon	Lebanese, Suaj, European breads
Libya	French bread, Arabic
Morocco	Moroccan Khobz El-daar, French, European
Oman	Arabic, Tanoor, Chapati, European
Qatar	Tanoor (naan), Arabic, European
Saudi Arabia	Samouli, Mafroud, Tannouri, Burr, Tamees, Korsan, European bread
Somalia	Injera
Sudan	Injera, Shamsi, Baladi
Syria	Mafrood (Arabic), Armani, Samoon, Suaj, European
Tunisia	Trabilsi, French bread
Turkey	Balzuma, Gomme, Yafka, French and other European bread
U.A. Emirates	Tanoor, Chapati, Arabic, Samoon, European
Yemen	Roti, Malouge, French bread, European bread

during which the sheeted dough aerates and its surface dries out, resulting in skin formation and subsequent puffing or the formation of double-layered product (Fig. 9-2) during baking. In the single-layered flat breads puffing and pocket formation are eliminated, either by the reduction of the final proofing period (prevention of skin formation) to 1 min or less or by baking the product at lower temperature for longer times. In addition, dough pieces of single-layered flat breads are usually docked or grooved prior to baking. This serves as a product ornament, and is another technique for preventing pocket formation by allowing the release of steam and CO_2 from the product. Figure 9-3 shows the processing steps of single- and double-layered flat breads.

Double-Layered Flat Breads

Arabic bread is the name most commonly given to double-layered flat breads popular in the Middle East and North Africa. Lebanese, *Mafrood*, *Pita*, and *Shami* are among other terms for essentially the same product. For the production of Arabic bread, the ingredients, flour (75–80% extraction), yeast (0.5–1.0%), salt (0.75–1.5%), and water are mixed. This results in a relatively stiff dough with a dough consistency of 650–850 BU (farinograph).

After mixing, the dough is allowed to ferment for an hour. This stage of processing sometimes is reduced to 30 min in commercial bakeries. Qarooni et al. (1989) found that extending the bulk fermentation time from 30 to 90 min had a significant ($p < 0.001$) effect on instrumental assessment of crust

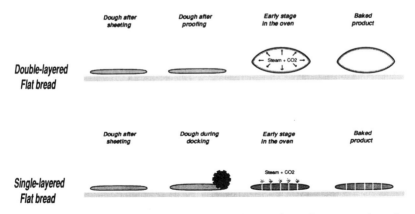

Figure 9-2. Pocket formation of double-layered, and suppressing the layer separation in single-layered flat bread.

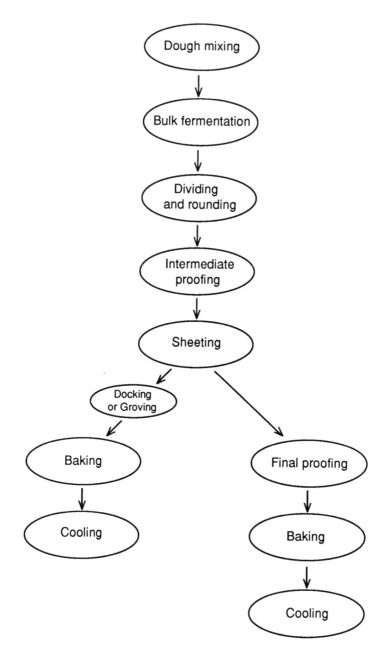

Single-layered Flat Bread   **Double-layered Flat Bread**

Figure 9-3. Processing steps of single- and double-layered flat bread.

and crumb color, the level of blistering, evenness of the two layers, and the grain appearance of Arabic bread. After bulk fermentation, the dough is divided, rounded, and allowed to proof for 10–15 min. In automated bakeries, the temperature and humidity of the proofing cabinet are adjusted to allow sufficient dough relaxation and to prevent either stickiness or skin formation. The desirable, soft, moist, and uniform crumb texture of Arabic bread is positively influenced by the dough temperature and intermediate proof time (Qarooni et al., 1989).

After the dough pieces have passed through intermediate proofing they are shaped to flat, round or oval pieces prior to final proofing. In automated bakeries, the dough pieces are passed under pressing rollers, to form a flat dough of 2–2.5 cm thick and then through two stage of gauge rolls which flatten the dough piece into round shape of desired thickness. In semi-automated bakeries, a small sheeter with two pairs of sheeting rolls is used. The dough pieces are manually transferred from the first to the second sheeting rolls. The thickness of sheeted dough ranges from 1.5 to 10 mm, based on consumer preference. Dough sheeting is the single most important processing step in the production of Arabic bread. Each double-layered flat bread has a narrow range of dough thickness that results in optimum product quality and small variations in thickness significantly influence that quality (Rubenthaler and Faridi, 1982; Qarooni et al., 1987; Williams et al., 1988; Quail et al., 1990).

After sheeting, the dough pieces are allowed to aerate again and relax prior to baking. In automated bakeries, the second proofing stage is done in a temperature- and humidity-controlled proofing cabinet. Several conveyor belts are installed in the cabinet on which the sheeted dough pieces are transferred along the top conveyor section and then turned over onto a belt below it. Hence, the final proofing time is controlled by the number of belts and their speed.

Baking of Arabic bread is carried out in high temperature ovens for a short time. A baking temperature of 470–500°C for 60–90 sec is reported for *shamy* bread (Doerry, 1983), 500–525°C for 45–60 sec for Mafrood (Mousa and Al-Mohizea, 1987), and 400°C for 90 sec for Arabic bread (Qarooni et al., 1987). The baking time and temperature vary depending on the thickness of the sheeted dough. Thin product is usually baked at high temperature (650°C) for a short time (18–20 sec). Although thin Arabic bread is desirable in many countries of the Middle East (Williams et al., 1988; Qarooni, 1989), maintaining the temperature of a large commercial ovens to such a high level is extremely difficult.

During the baking of double-layered flat breads (Arabic and *Baladi*), the thin skin of sheeted dough that was formed during the final proofing stage changes to a pale colored crust, and the internal temperature of the central portion of dough rises to 99°C (El-Samahy and Tsen, 1983), causing the steam to develop. The combined action of steam pressure and that of CO_2 causes the separation of top and bottom layers. In all thicknesses of sheeted dough and baking conditions used by Quail et al., (1990) the pocket formation occurred during the first third of the baking time.

After baking, bread is cooled prior to packing. In automated bakeries, fine spray nozzles are used immediately after baking to apply water on the surface of bread. The water dries out, leaving an attractive glaze on the surface of the product. Breads are transferred (10–15 min) to the packaging area on a cooling conveyor. Before packaging, the interior bread temperature should be about 35–37°C. Insufficient cooling might give rise to many problems such as difficulties in separation of bread layers and excessive moisture condensation within the wrapping (Qarooni, 1989).

***Baladi* Bread.** The typical formula for *Baladi* bread (Doerry, 1983) contains flour of 82% extraction (100%), water (70–75%), salt (0.5%), and sour dough from a previous batch (20%). A typical starter consists of an old dough (5.5 Kg), flour (50 Kg), and water (25 Kg). A minimum of 3 hours of starter fermentation is allowed to take place. Compressed yeast is added to the starter and sometimes to individual doughs during the winter months to compensate for the lower dough temperature. The ingredients are mixed for 15–25 min. Due to its high water content, the dough is very slack. This makes mechanical processing of the product difficult but, has a significant positive impact on the physical characteristics and flavor of the bread (Dalby, 1963; Mousa et al., 1979). The dough (27–28°C) is fermented for 40 min and divided into 180 g pieces, which are formed into a balls and placed on a wooden tray covered with a thin layer of bran. An initial proof time of 15 min is allowed after which, the dough is flattened by hand into a round shape of 20 cm diameter and 1.25 cm thickness. Final proof is 50–60 min. Prior to baking, the dough pieces are hand pressed further to make the desired diameter, excessive bran is removed, the dough pieces (5–6) placed on a long wooden peel, transferred to the hearth of the oil-fired oven, and baked at 450–600°C for 90 sec (Dalby, 1963). Baking periods of 2–3 min at 350°C (Hamed et al., 1973), and 3–4 min at 300–350°C (Finney et al., 1980 and Mousa et al., 1979) have been reported. Sometimes freshly baked *Baladi* bread is brushed with water and placed in another oven (200°C) for a few minutes (Dalby, 1963) to produce a toasted product.

Single-Layered Flat Breads

Tanoor **bread.** Similar to Arabic bread, a choice of common name for this product is difficult. Several names, *Tanoori, Tandour, Khubz, Naan* are given to essentially the same product in various parts of the Middle East. In Iran, *Tanoor* (locally called *Taftoon*) bread (Fig. 9-4) is made from flour (100%) of 84% extraction, soda (1%), sour dough (50%) or yeast (0.5%), water (60%), and as an optional ingredient, date syrup (2.5%) (Faridi and Finney, 1980). However, the addition of date syrup is infrequent due to its high price. The formula of *Tanoor* bread used in seven commercial bakeries in Kuwait (Qarooni, 1989) consisted of flour of 90% extraction (100%), soda (0.0–0.3%) salt (0.7% in winter and 1–1.5% in summer), active dry yeast (0.1%), and water (65%). Soda is added to improve dough handling and to develop a desirable crust color. Because of availability of active dry yeast in the local market, use of sour

Figure 9-4. *Tanoor* bread being baked in a clay oven.

dough is now less popular. The ingredients are mixed for 15–20 min in a wishbone mixer, the dough allowed to ferment for 1–2 hr, divided into 220-g pieces, rounded by hand, and given an intermediate proof time of 10–15 min. In dry weather, the dough pieces are covered with a cloth to prevent skin formation. Each dough piece is then sheeted by hand or with a roller to the proper thickness. It is docked, spread on a special cushion, and immediately stuck to the wall of a clay oven. Docking prevents pocket formation and promotes uniform baking to be achieved. Baking is carried out for 45–120 sec depending upon the distance of the sheeted dough from the heat source. The bread peels off the oven wall and it is immediately removed.

Barbari bread is most popular in the Northern part of Iran. Usually 70–80 cm long and 25–30 cm wide with a thickness of 2.5–3.0 cm (Fig. 9-5), it is thicker and heavier than other Iranian breads. A typical formula (Faridi and Finney, 1980) includes wheat flour of 77% extraction, (100%), soda (0.35%), salt (2.0%), sour dough (40%) or ycast (1%), and water (~60%). The ingredients are mixed to the desired consistency and fermented for 2 hr. Dough balls of 800–900 g are formed, allowed to rest (intermediate proof) on a table, then flattened to oval shapes and rested for 20 min. One or two tablespoon of "Roomal," a gelatinized flour:water (10% w/v) paste is poured on each flattened dough piece. The paste is spread evenly on the sheeted dough with fingers, while forming a number of grooves (2–3 cm apart) along its length. Grooving prevents the separation of the top and bottom crusts and also decorates the loaf. Sheeted dough is lifted with both hands and placed on the peel which is used to transfer it to the oven. During lifting, the sheeted dough is further stretched to its final length prior to baking. The dough is baked at 220°C for 8–12 min on the hearth of a brick oven (Faridi and Finney, 1980).

Korsan. Single-layered, circular (57 cm), thin, and flat *Korsan* should have light brown crust color. The ingredients, whole meal flour (100%), salt (0.35%), yeast, and water are mixed manually for 30 min, allowed to ferment (30 min), and remixed for a second time (15 min). After final fermentation, (60 min) the dough is manually divided (180–200 g) and allowed to proof for 30–45 min. It is then sheeted by hand and baked at 150–200°C for 2–3 min (Mousa and Al-Mohizea, 1987).

Lavash is an oval-rectangular (60–70 cm, 30–40 cm), flat bread (220 g) of 2–3 mm thickness. Its creamy-white surface should be covered evenly with small blisters. Flour of 82% extraction (100%), water (~45%), salt (2.0%), and soda (0.25%) are mixed to a dough which ferments for 1–3 h. Dough balls of 300 g proof for 5–10 min before being sheeted to thin layers with a roller on a wooden board. Sheeted dough is placed on a special cushion to be stuck to the wall of the oven. The bread dries out quickly during and after baking and can be stored for 3–6 months under proper conditions (Faridi and Finney, 1980).

Moroccan Wheat Bread. This round bread is made in two stages. Portions of white flour (70%), water (~45%), and yeast (1%) are mixed and allowed to ferment for 90 min. Then salt (2%), the remaining flour (30% whole meal), and water (~15%) are added and mixed to optimum development. The dough is fermented for an additional 30 min, rounded, flattened to a

Figure 9-5. *Barbari* bread.

disk 18 cm in diameter and 5 cm thick, and proofed for 45 min. The proofed dough is baked for 20 min at 220°C (Faridi, 1988).

Naan. The ingredients are flour (100), yeast (2%), sugar (1%), salt (2%), water (~35%), yogurt (25%), and shortening (6%). It is also prepared from the four essential ingredients of leavened bread, i.e., flour, water, salt, and yeast or sour dough. The dough is mixed, immediately divided into balls, and allowed to ferment while covered with a wet cloth. It is then sheeted into thin, oblong, flat pieces and baked at 315°C for 2 min (Faridi, 1988).

Sangak. This sour dough bread's name drives from a Farsi term for "small stone." The loaf is triangular in shape and flat (70–80 cm long, 40–50 cm wide with 3–5 mm thickness). The bottom crust is full of indented blisters because of its direct contact with hot pebbles during baking. The top crust, also possessing many small blisters is usually sprinkled with sesame or poppy seeds. Dough ingredients include 87% extraction flour (100%), water (~85%), sour dough (20%), and salt (1%). All ingredients are mixed and then fermented for 2 hr. A portion of the resulting soft dough (500 g) is sheeted on a special convex paddle, docked, and transferred onto hot pebbles covering the oven floor. The pebbles are moistened from time to time with a soap solution to ensure easy removal of the bread (Faridi and Finney, 1980). The temperature of the pebbles varies between 350 to 500°C and the baking time between 2 to 4 min (Maleki, 1984).

Shamsy. This disk-shaped bread (500 g and 20 cm, diameter) with a brownish crust and white, firm crumb is made from a lean formula, consisting of flour, water, salt, and sour dough (old dough, 2 Kg flour, and 0.5 liter water). This is hand mixed for 25 min and fermented for 45 min. Dough pieces are then covered with wheat bran and left in a sunny place for 60 min for a second fermentation. They are pressed by hand to form a flat dough of 20 cm diameter and left in the sun for a third 60 min after which are turned over and left in the shade for a final 30 min fermentation. They are baked in special a homemade oven (El-Gendy, 1983).

Terabelsi. To produce *Terabelsi*, a round bread of 20 cm diameter, white flour (100%), water (~60%), salt (2%), and yeast (1%) are mixed to optimum development and fermented for 30–45 min. The dough is divided into 700-g balls, rolled out to a thickness of 2 cm then proofed for 45 min. Four cuts are made across the top to form a square before the dough is sent to the oven for baking at 220°C (Faridi, 1988).

Ovens Used for Flat Breads

The baking industry in the Middle East and North Africa uses a wide range of oven designs. The traditional ovens for flat bread baking, a barrel shaped oven or heated surface (Adler, 1958; Faridi and Finney, 1980; Pomeranz, 1987), similar to those used by the ancient Egyptians, are still widely used. The barrel shaped ovens are installed, either vertically in or on the ground or fixed at a 45° angle to the floor of the bakery (Fig. 9-6). In larger bakeries, two or three of these ovens are installed (at a 45° angle to the floor). The baker usually loads the inner wall of the first oven and moves to the second one. By the time the second oven is loaded, the baking in the first oven is complete. Traditionally, these ovens were heated with wood or with dry leaves of palm trees (in date palm growing areas). However, because of the convenience of oil burners, and the high production demand before breakfast, lunch, and dinner, the use of wood has diminished, except, in the remote rural areas.

Barbari bread in Iran is baked in short, dome-shaped (75 cm high and 350 cm diameter), brick ovens, heated with a petrol burner from the side. *Sangak* (Faridi and Finney, 1980) and *Tarboun* (Adler, 1958) breads are baked on heated pebbles. The oven for *Sangak* bread is made of sun-dried clay bricks (4 × 4 m). The hearth (which is inclined toward the inlet of the

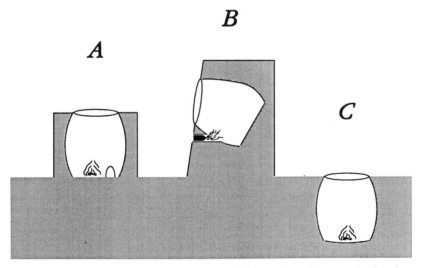

Figure 9-6. Variations in installation of *Tanoor* bread oven, **A.** On the ground; **B.** 45° angle to the ground; and **C.** In the ground.

oven) is 1.5 m above the bakery floor and covered with small pebbles (7.5 mm in diameter). The oven is heated with a petrol burner from a corner of the oven. The flame is isolated from the baking area by a 1 m high brick wall. To ensure easy removal of baked product from the heated pebbles, occasionally their surface is coated with a soap solution. Figure 9-7 shows

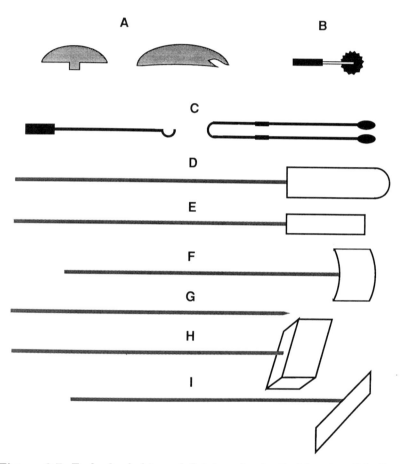

Figure 9-7. Tools for baking of flat breads. **A.** Cushion to stick the sheeted dough of the wall of the oven; **B.** Docker; **C.** Tools to remove the baked product from the wall of the oven; **D.** Wooden peel to transfer the dough of Arabic and Baladi bread to the oven; **E.** Metal peel to remove the flat breads (Arabic and Baladi) from the oven; **F.** Concave metal peel to form the dough of Sangak bread to thin layer and transfer it to the oven; **G.** Metal rod to separate the Sangak bread from the pebbles; **H.** Heavy wooden block to level the pebbles in the oven of Sangak bread making; **I.** Tool to distribute the pebbles in the oven of Sangak bread making.

some of the tools commonly used in the traditional method of flat bread making.

Automated production lines for flat bread (Arabic, *Barbari*, *Lavash*, etc.) use gas fired tunnel ovens with a traveling hearth. For thin (2 mm) Arabic bread, ovens capable of maintaining 550°C or higher are required.

Semolina Products

Semolina has a number of food uses in the Middle East and North Africa. In North Africa, it is used for the production of the national dish, *couscous*. A type of flat bread is prepared from a mixture of wheat flour and semolina (*Matlowa*, in Algeria), as well as various types of pasta. In the cities of the Middle East, pasta is consumed, mainly in the form of elbow macaroni, short spaghetti, and nest. Semolina is also used in the production of a number of cakes and sweet goods.

Couscous is the name given to a processed semolina which is a major food staple in North Africa (including Algeria, Libya, Morocco, and Tunisia) where it is consumed two to three times a week. Other terms for essentially the same product include *couscousi, couscouso,* and *aesh.* In North Africa, couscous is made from durum wheat semolina, but other grains may be used (Williams, 1981). The term couscous is locally used for the popular stew dish prepared and served with steamed couscous. The preparation of this meal is done in a couscousiere, a vessel which consist of two pans, a small one that has a perforated base and fits tightly on the large one. Dry couscous is prepared at home to provide a supply for the entire year. To ensure a low moisture content, it is dried under the sun during hot summer months (Kaup and Walker, 1986).

The traditional procedure of couscous making (Fig. 9-8) (Williams, 1981; Kaup and Walker, 1986) includes moistening of semolina in a wide wooden container, mixing and rubbing the bulk with the palm of the hand (agglomeration) to form small granules, steaming, and drying. For couscous of uniform particle size, one or more sieve separations are used. The bulk of fine particles are transferred back to the wooden container to be treated in the same manner until the necessary amount has been prepared. The agglomerated semolina is transferred to the upper pan of couscousiere and steamed. After steaming, the coarse agglomerates are sun dried and stored.

According to Quaglia (1988) the first large-scale (500 Kg/h) production line of couscous was established in 1979, in Safax, Tunisia. Commercial line processing of couscous follows the

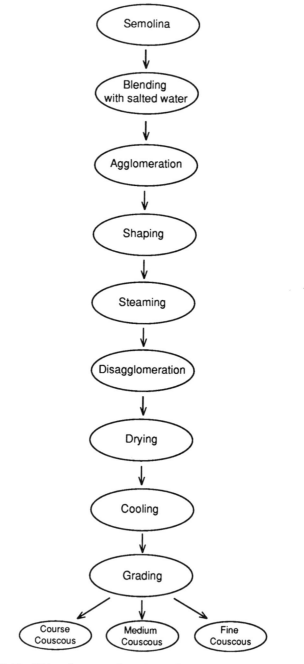

Figure 9-8. Traditional processing steps of couscous.

traditional procedure. The raw materials (semolina and water) are mixed to form a mass of 30–40% moisture. Mixer geometry and speed plays an important role in production of a homogeneous material. Coarse materials leaving the mixer are transferred to a detacher for size reduction. The agglomerated material is cooked with steam at 120°C for 4 min. The extent of starch gelatinization at this stage depends on the size and moisture content of agglomerates and their cooking time. After steaming, the material is sent to a separator to disagglomerate the large lumps before drying. Precooked materials are dried on a conveying belt to the safe moisture content of 10–12%. Couscous is graded to fine, medium, and coarse products through sifting. The fine material is recycled, and the coarse is reduced by a roller mill before packaging.

Bulgur

Bulgur is a term commonly used in the U.S. for an ancient wheat-based product known to have originated in the Middle East and North Africa where it is often referred to as *burghoul*, *balgour*, and *boulgur*. Bulgur is still made by relatively primitive methods in Lebanon, Turkey, Iran, Iraq, and India. Pre-soaked wheat grain is boiled with excess water (Fig. 9-9). During boiling, buoyant foreign materials float to the surface and are removed. Cooked wheat is dried under the sun after which time the outer bran layers are removed by abrasion, and bulgur is produced by cracking the resultant product into coarse pieces (Haley, 1960; Edwards, 1964).

Due to gelatinization of starch, Bulgur's texture becomes glassy, hard, and, hence, more resistant to insect damage. It is prepared usually after wheat harvest and can be stored under a wide range of temperature and humidity conditions (Neufeld et al., 1957). Durum wheat is preferred over hard and soft wheats for bulgur making in the Middle East, because of the color and hard texture of endosperm (Williams et al., 1984).

A mechanized process for Bulgur production was described by Smith et al. (1964). The processing involves soaking and pressure cooking of cleaned wheat in batch type rotary units. The total soaking and cooking time is about 2 hr. The resultant product is dried in a column-type grain dryer. A continuous process was also developed and patented by Robbins (1959). Cleaned wheat is heated and soaked to a final moisture of 45%. This is achieved through three large tanks connected in series. Total soaking time in this procedure is 8 hr. Soaked

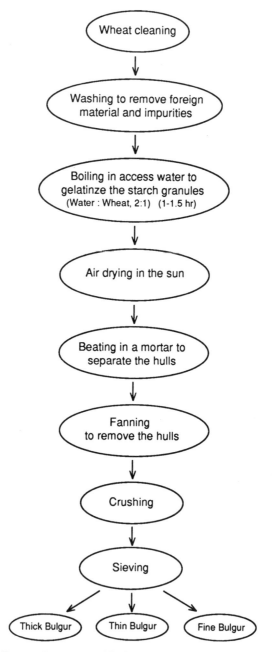

Figure 9-9. Processing steps of Bulgur.

wheat is cooked in a continuous pressurized cooker for about 1.5 min, and the resultant gelatinized wheat kernels are dried in two stages: first, by passage through a rotary drier to remove the surface moisture; and second, through a column-type drier to reduce the wheat moisture to 10–11%.

Cakes, Pastries, and Sweet Products

Although flour quality dictates the properties of the end products, the baking industry of the most countries in the Middle East and North Africa has access to limited types of wheat and flour. Therefore, a wide range of traditional and European biscuits (cookies), pastries, and cakes are made from existing flour types. This section describes the processing methods for various types of traditional Middle Eastern and North African pastries, cakes, cookies, and specialty products such as *Kenafeh*.

Bameah. Flour, water, oil or melted butter, and eggs constitute the basic formula. The mixture of oil and water is heated to boiling. Flour is added and mixed for a short time. When the mixture is cool the whole eggs are added to form a smooth paste. Small portions of this mixture are deep fried in cooking oil. *Bameah*, the golden brown product (Fig. 9-10) is cooled and

Figure 9-10. *Bameah.*

dipped in a thin syrup of water, sugar, lemon juice, and flavoring.

Drabeel, Karapeech, or ***Youkeh*** all refer to a product prepared from flour of high extraction (80–85%) which is mixed with water and a small amount of oil to form a cohesive mass. After a short rest time, a portion of this dough is sheeted to about 1 mm thickness with a number of wooden roller pins. The dough is then transferred to a concave metal griddle, heated by a burner from below. After a short baking time and while the baked dough is still flexible, a portion of finely ground sugar, cardamon, and cinnamon is spread in the middle, and the sheeted dough is folded or rolled over to form a long rectangular or roll-shaped product. At this stage, the product is baked on the griddle to optimum color. The baked product, while on the griddle, is cut to uniform size, and allowed to cool prior to packaging.

Halwa or ***Halva.*** A series of products are produced under a general term of *halwa* (Arabic word for sweet). One kind of *halwa*, reported by Sawaya et al. (1985), is prepared from dehulled roasted sesame seed, a mixture of glucose and sucrose or sucrose alone, citric acid, and *halva* root extract obtained from *Saponaria officinalis*, which is normally used for its antifoaming, whitening, and emulsifying-stabilizing properties. Flour *halwa* consists of flour (in many cases 85–90% extraction), sugar, and small amounts of water (or rose water), citric acid (lemon juice), and flavoring materials such as saffron and cardamon. The processing of these products starts with preparation of syrup, which consist of sugar, water, small amount of lemon juice, cardamom, and saffron. The mixture is heated to form a thin syrup. Flour and oil are placed in a frying pan and continuously mixed on a moderate heat to form a light-brown paste. While hot, a portion of the syrup is added to the flour-oil paste. This step presumably allows the oil-coated flour particles to hydrate and form a cohesive mass. The consistency of the final product is controlled by the amount of added syrup. Sometimes a portion of various kinds of ground nuts is added.

Homemade-Style Cookies. Flour of moderate (75%) to high (80–90%) extraction rate is used to produce a range of homemade-style cookies. The basic formula for all these products consists of flour (100%), sugar (25–25%), shortening (35%), and water. In many instances, rose water substitutes for part of the water. Finely ground green cardamon or a combination of ground cumin and fennel seeds is used as flavoring. In the latter case, very little sugar is included in the formula, and the product is consumed with a mixture of finely ground sugar, sesame seeds, and cinnamon. After combining all the ingredients, the homogeneous short dough mass is allowed to rest for an hour, divided into pieces and placed in a wooden die, smaller, but similar to, that for Scottish shortcakes which leaves a special design on the product. The embossed dough pieces then are baked. In many instances, the dough pieces are stuffed with dates, nuts, and spices to produce stuffed cookies (Fig. 9-11).

Kenafeh. Similar to a soft, thin noodle, *Kenafeh* is the essential ingredient of a number of sweet goods. The ingredients (wheat flour, corn starch, milk powder, oil, and water) are thoroughly mixed to form a thin batter which is transferred to a depositing tank, with multiple nozzles. Thin streams of the batter are deposited on a rotating heated plate (Fig. 9-12) for a short cooking time. The baked strips of *kenafeh* are collected and used for the production of a number of sweet products, such as cheese *kenafeh*.

Legaymat is a traditional Middle Eastern product prepared from a mixture of high extraction flour (80–85%) and starch (2:1), a small amount of oil, baking powder or yeast, yogurt, and water. The ingredients are mixed to form a thick batter, flavored by ground cardamon and saffron. The batter ferments

Figure 9-11. Home-made style cookies.

for 4 to 5 hours. Spoonful portions of this batter are deep fried to produce the final ball-shaped golden brown product. This is cooled and dipped into a thick syrup of water, sugar, lemon juice, finely ground cardamon, and saffron.

Frekeh is a product prepared from immature wheat (Williams and El-Haramain, 1985, cited by Dick and Matsuo, 1988). Wheat is cut at the stem and allowed to dry partially under the sun. It is bunched together and scorched (parched) by burning off the awns and leaves. The flavor of *Frekeh* derives from this scorching process.

Kishk or *kashk* is a dried mixture of ground wheat and heavily fermented dairy products. It consists of small, round or

Figure 9-12. Production of *Kenafeah*. Deposit of batter strips on rotary cooking plate.

irregular pieces with a yellowish-brown color, rough surface and hard texture. When soaked in water, it changes to the creamy-white paste which is added to many types of meals in the Middle East and North Africa, for taste and flavor improvements. In Egypt (El-Gendy, 1983), *Kishk* is prepared from *laban zeer* (primarily a thick sour buttermilk) or full cream sour milk and boiled, dried, ground wheat. The process of making *kishk* is usually carried out during July and August of each year when the temperature is highest in Upper Egypt. Wheat is boiled in water and dried under the sun. It is crushed to fine particles in a stone mill and the bran is removed by sieving. A small amount of ground wheat is moistened with salted water then mixed with *laban zeer* (diluted with raw milk or water) to form a homogeneous, thick paste. The mixture is fermented for a minimum of 24 hours then thoroughly mixed, divided, made into small round shapes, and dried for several (2–3) days under the sun. A small amount of spices might be added to the mixture before drying.

Harees, Haleem (Pearled Wheat). Pearled wheat is used to prepare a traditional dish in many parts of the Middle East. Wheat kernels are moistened and then bran layers are removed by stone mills or wooden mortar. A large scale scourer or pearler can be used for the process. This product is used to produce porridge-like product known as "Harees" or "Haleem." For the preparation of a *harees* dish, beef, turkey, or chicken meat is placed with a mixture of *harees* and water (1:1) in a covered cooking pot and allowed to cook over a moderate heat for 4 to 5 hours. After cooking the liquid is separated from the cooked meat and pearled wheat. The latter mixture is blended and mashed to form a homogeneous paste. From time to time, part of the separated liquid is added to ease the process of pasting. A small amount of salt is added. *Harees* is served with ground cinnamon and sugar. This product is usually prepared at home or in small shops for breakfast or dinner. The most important aspect of *harees* making is the homogeneity of the final product. Color and viscosity of the final product are of significant importance to its acceptability.

Roasted Wheat. Traditionally, roasted wheat products are prepared by boiling whole wheat kernel in excess water. In the last stage of cooking, a small amount of salt is added to the mixture and cooking continues for a couple of minutes. After draining, the cooked wheat is dried under the sun. The gelatinized and dried wheat kernels are dry roasted for use as a snack food.

Future Outlook

The population growth in the countries of the Middle East and North Africa resulted in significant migration of manpower to the cities, to work and a better life. This has led to over populated cities and a responsibility for the local governments to provide the staple foods. Large automated and semi-automated milling and baking plants were established by the governments and private sector to fulfill this task. For the encouragement of these industries a significant effort should be made to adapt the traditional products to today's manufacturing standards. Unfortunately, apart from a limited number of publications on the most popular wheat based products, very little has been done to study other products. Research on these products has a double benefit. First it improves the quality of local products and nourishes the industry, and second it allows development of new products in other countries of the world.

Acknowledgment

Contribution No. 93-23 from the Kansas State Agricultural Experiment Station, Manhattan, KS 66506-2201

References

Adler, L. 1958. Breadmaking in the land of the Bible. Cereal Sci. Today 3:28-30, 32.

Al-Mashhadi, A., Naeem, M., and Bashour, I. 1989. Effect of fermentation on yield and quality of irrigated Yecora Rojo wheat grown in Saudi Arabia. Cereal Chem. 66:1-3.

Al-Hazek, M. T. 1988. Econometric analysis of the wheat problem in Egypt. J. Agric. sci., Alexandria Univ. Egypt. 13:802-822. (abstract).

Amr, A. S. 1988. Effect of growing season, location, and variety on the quality of the commercially-grown durum wheats in Jordan. Dept. Food and Technology, University of Jordan. Amman, Jordan (abstract).

Anonymous. 1990. Saudi aim to cut wheat exports. Agric. outlook. April.

Bahram, N. 1986. Ecological boundary for wheat cultivation in Jordan. University of Jordan. Amman, Jordan (abstract).

Bakhella, M., Hoseney, R. C., and Lookhart, G. L. 1990. Hardness of Moroccan wheat. Cereal Chem. 67:246-250.

Briggle, L. W., and Curtis, B. C. 1987. Wheat worldwide. in Wheat and Wheat Improvement, Heyne, E. G. ed. American society of Agronomy, Inc., Crop Science Society of America, Inc., and Soil Science of America, Inc. Publishers. Madison, Wisconsin, USA. p.25.

CIMMYT (Center International de Mejoramiento de Maiz y Trigo). 1987-88. Third world wheat supply and demand to the year 2000. In world wheat facts and trends. p. 20-24.

Chazan, M. and Lehner, M. 1990. An ancient analogy: Pot baked bread in ancient Egypt and Mesopotamia. Paleorient 16:21-35.

Choudhri, M.B., Bashir Choudhri, I.M., Bunting, A.H. and Bunting, E.J., 1987. Food and Agriculture Organization of the United Nations (FAO), Rome, Italy.

Dalby, G. 1963. The baking industry in Egypt. Baker's Dig. 37 (6):74-77.

Demirlicakmak, A., Ergin, Y., and Yakar, K. 1972. Results of third IWWPN grown at Ankara, Turkey. Proc. 1st Winter Wheat Conf. Ankara, Turkey, p. 259.

Demirlicakmak , A. 1972. Problems and opportunities for increased winter wheat production in Turkey. Proc. 1st Int. Winter Wheat Con., Ankara, Turkey, p. 3.

Dick, J. W. and Matsuo, R. R. 1988. Durum wheat and pasta products. In: Wheat Chemistry and Technology. Pomeranz, Y. ed., 3rd ed. Volume II, American Association of Cereal Chemist, Inc. St. Paul, Minnesota, p. 507-547.

Doerry, W. 1983. Baking in Egypt. Cereal Foods World 28:677-679.

Edwards, G. H. 1964. Bulgur, Ala of American rice. Milling, April, 3, The Research Association of British Flour Millers, St. Albans, England.

El-Gendy, S. M. 1983. Fermented foods of Egypt and the Middle East. J. Food protection. 46:358-367.

Elings, A., and Nachit, M. M. 1991. Durum wheat landrace from Syria, Euphytica. ICARDA 53 (3):211-224.

El-Samahy, S. K., and Tsen, C. C., 1983. Changes of temperature inside the loaf of Baladi and Pan breads during baking. Chem. Technol. Lebensm. 8:15-18.

Faridi, H. A. 1988. Flat bread. In: Wheat Chemistry and Technology, pages 457-498. ed. Y. Pomeranz. Vol II 3rd ed. Am. Assoc. Cereal Chem., St. Paul, Minnesota, USA.

Faridi, H.A., Finney, P.L. and Rubenthaler, G.L., 1981. Micro baking evaluation of some U.S wheat classes for suitability in Iranian breads, Cereal Chem., 58:428-432.

Faridi, H. A., Finney, P. L. 1980. Technical and nutritional aspects of Iranian breads. Baker's Dig. 54(5):14-22.

Finney, P. L., Morad, M. M., and Habbard, J. D. 1980. Germinated and ungerminated faba bean in conventional U.S. bread made with and without sugar and Egyptian Balady breads. Cereal Chem. 57:267-270.

Food and Agriculture Organization of the United Nation (FAO). 1976. Production yearbook, 29:60-61. Rome, Italy.

Hafiz, A. 1972. Winter wheat improvement problems in the Near East and North Africa. Proc. 1st. Winter Wheat Conf. Page 22-24. Ankara, Turkey.

Haley, W. E. 1960. Bulgor, an ancient wheat food. Cereal Sci. Today. 5(7):203-207, 214.

Hamed, M. G. E., Rafai, F. Y., Hussein, M. F., and El-Samahy, S. K.

1973. Effect of adding sweet potato flour to wheat flour on physical dough properties and baking. Cereal Chem. 50:140-146.

ICARDA (International Center for Agricultural Research in the Dry Area). 1983. Better harvests in dry areas. Aleppo, Syria.

Kaup, S. M., and Walker, C. E. 1986. Couscous in North Africa. Cereal Foods World 31:179-182.

Khatchadourian, H. A., Sawaya, W. N., and Bayoumi, M. I. 1985. The chemical composition and rheological properties of flours milled from two major wheat varieties grown in Saudi Arabia. Cereal Chem. 62:416-418.

Malik, M. A. K., El-Tayeb, M. B., and Ouden, H. A. 1981. Agricultural production versus industrial capacity in Libya. Third report. National Academy for Scientific Research, 5-32. Tripoli, Libya. (Abstract).

Maleki, M. 1984. The characteristics of Iranian wheat and flour and the technology of Iranian Bread. University of Shiraz, p. 8 (in Farsi).

Mohammed, J. 1988. Increasing wheat production in Ethiopia by introducing double cropping under irrigation. The Fifth Regional Wheat Workshop for Easter, Central and Southern Africa and Indian Ocean. Canadian International Development Agency, Ottawa: CIMMYT, 161-165.

Mousa E.I., Ibrahim, R.H., Shuey, W.C., and Maneval, R.D. 1979. Influence of wheat classes, flour extraction, and baking methods on Egyptian Balady bread. Cereal Chem. 56:563-566.

Mousa, E. I., and Al-Mohizea, I. S. 1987. Bread baking in Saudi Arabia. Cereal Foods World, 32(9):614-620.

Nagao, S. 1981. Soft wheat uses in the Orient. Page 276. in "Soft Wheat: Production, Breeding, Milling, and Uses." Page 276. Yamazaki, W. T., and Greenwood, G. T. ed., Am. Asso. Cereal Chem. St. Paul, Minnesota.

Neufeld, C. H. H., Weinstein, N. E., and Mecham, D. K. 1957. Studies on the preparation and keeping quality of bulgur. Cereal Chem. 34:360-370.

Patel, K. M., and Johnson, J. A. 1975. Horsebean protein supplements in breadmaking. III. Effect of physical dough properties, baking and amino acid composition. Cereal Chem. 52:791-800.

Pomeranz, Y. 1987. Modern Cereal Science and Technology. VCH publishers. p.259.

Qarooni, J. 1989. Handbook of Arabic bread production. Kuwait Flour Mills and Bakeries Co. Kuwait. p.25.

Qarooni, J., Miskelly, D., and Wootton, M. 1989. Factors affecting the quality of Arabic bread - Fermentation variables. J. Sci. Food Agric. 48:99-109.

Qarooni, J., Orth, R. A., and Wootton, M. 1987. A test baking technique for Arabic bread quality. J. Cereal Sci. 6:69-80.

Quail, K. J., McMaster, G. J., Tomlinson, J. D., and Wootton, M. 1990. Effect of baking temperature/time conditions and dough thickness on Arabic bread quality. J. Sci. Food Agric. 53:527-540.

Quagilia, G. B. 1988. Other durum wheat products. In Durum Chemistry and Technology., ed Fabriani, G. and Lintas, C., p. 253-282. Am. Assoc. Cereal Chem. St. Paul, Minnesota.

Robbins, D. H. 1959. Method of processing wheat. U.S Patent 2,884,327.

Rubenthaler, G. L., and Faridi, H. A. 1982. Laboratory dough moulder for flat breads. Cereal Chem. 59:72-73.

Sawaya, W. N., Khalil, J. K., Ayaz, M., and Al-Mohammad, M. A. 1985. Chemical composition and nutritional value of Halva. Nutrition Reports International. 31:389-397.

Seeborg, E. F. 1974. Turkey's wheat outlook clouded by weather. For. Agric. 12 (25):2.

Sjerven, J. 1989. Saudi Arabia's HADCO: wheat farmer and more. Agribusiness Worldwide. 11 (8):10-15.

Smith, G. S., Barta, E. J., and Lazar, M. E. 1964. Bulgur production continuous atmospheric-pressure process. Food Technology 89-92.

Storck, J., and Teague, W. D. 1952. Flour for man's bread. Pages 34, 64. University of Minnesota Press. Minneapolis.

US Wheat Associates, Inc. FY 1991-92.ᵃ Marketing plan, Middle East-East Africa. Pages I1-I37, H1-H63. Headquarter Office, Washington, D.C.

US Wheat Associates, Inc. FY 1986-87. Marketing plan, Middle East-East Africa. Pages H1-H55. Headquarter Office, Washington, D.C.

Williams, P. C., El-Haramein, F. J., Nelson, W. and Srivastava, J. P. 1988. Evaluation of wheat quality for the Baking of Syrian-type two-layered flat breads. J. Cereal Sci. 7:195-207.

Williams, P. C. 1981. Preliminary reports on grain utilization as food in countries served by ICARDA. Visit to Tunisia, March 1-4.

Williams, P. C., El-Haramein, F. J., and Adleh, B. 1984. Bulgur and its preparation. Rachis 3(2):28-30.

Williams, P. C., and El-Haramein, F. J. 1985. Frekeh making in Syria. A small but significant local industry. Rachis 4(1):25-27.

World Grain. 1985. Focus on Saudi Arabia. 3:17-18. Sosland Publishing Co. Kansas City. Mo.

World Grain. 1988. Focus on Turkey. 7:28. Sosland Publishing Co. Kansas City. Mo.

World Grain. 1987. Focus on Turkey. 5:27-28. Sosland Publishing Co. Kansas City. Mo.

World Grain 1987. Focus on Morocco. 5:31-32. Sosland Publishing Co. Kansas City. Mo.

World Grain. 1991. Focus on Israel. 6:20. Sosland Publishing Co. Kansas City. Mo.

Yamazaki, W. T., Ford, M., Kingswood, K. W., and Greenwood, C. T. 1981. Soft wheat production, pages 6-7 in "Soft Wheat: Production, Breeding, Milling and Uses," Yamazaki, W. T., and Greenwood, G. T. ed., Am. Asso. Cereal Chem. St. Paul, Minnesota.

CHAPTER 10

Wheat Usage in Southern and Central Africa

Philip G. Randall, Arie Wessels, and Hans W. Traut

Introduction

Wheat is not indigenous to Southern and Central Africa. The climate, tropical and savanna, favors cereal crops such as maize, sorghum, millet and rice. Although South Africa, Zimbabwe and Kenya have succeeded in producing wheat commercially under irrigation, the vast majority of wheat is produced under dryland conditions (only 2% of the area under cultivation in Africa is irrigated). Only Zambia and Angola are fortunate to have suitable climates for wheat production.

Wheat was introduced into the Cape area of South Africa in the late 17th century, though it was not until the mid 18th century that production spread and larger quantities of wheat was produced (Anon, 1990). Colonial rule in several African countries, in the late 19th century, led to limited wheat production, for the use of the colonists and, more importantly, for the emerging African elite. The acquired taste for wheaten products was not lost during the 1960s and early 1970s, with the retreat of the colonial governments, and as rapid urbanization occurred the demand for bread has increased.

During this period most southern African countries were self-sufficient in food production and had economies sufficiently buoyant to import their wheat requirements. By the late 1970s and early 1980s the region was suffering an economic crisis and, despite assistance from the International Monetary Fund (IMF) and the World Bank, many economies have yet to stabilize. The reasons for the decline can be summarized as follows (FAO, 1992):

1. Declining trade.
2. A heavy, and accumulating debt burden.
3. Severe, and recurrent drought.
4. Political conflicts.
5. Population growth.
6. Shocks in the world economy.

Of the approximately 42 countries with a per capita national income of less than US $500 per year, some 31 of them are in Africa, particularly in East, Central and West Africa (World Bank Atlas, 1990). In 1989 only ten out of 50 African countries had a per capita national income exceeding US $1,000 (World Bank Atlas, 1990).

In the 1990s the International Wheat Council (IWC) reported the use of wheat is now primarily determined by financial constraints (balance of payments) rather than by demographic features (Anon, 1991a). This economic stress has obliged many countries to cut back on wheat (amongst other cereals) imports, some even going as far as imposing import bans *i.e.* Nigeria and Zimbabwe.

The drought during the late 1980s and up to date has been the worst in more than a century (Sanai, 1992) and has ravaged cereal crop production. The majority of people on the continent earn a living in agriculture which in many countries is largely subsistence farming (Esterhuysen, 1992). The *per capita* food production has declined during the 1980s (FAO, 1990) which gives cause for concern in view of the rapid popu-

Table 10-1. Wheat Production in Southern and Central Africa (1,000 t), 1984-1989[a]

Country	1984	1985	1986	1987	1988	1989
Angola	10	10	10	2	2	2
Kenya	145	279	270	207	243	258
Lesotho[b]	...[c]	...	...	3	2	3
Namibia[b]	...	...	...	6	5	5
Tanzania	74	83	72	75	76	97
Zambia	...	12	18	39	40	40
Zaire	21	20	20	25	35	35
Zimbabwe[b]	99	215	248	213	260	285

[a] Botswana, Malawi, Mozambique, Nigeria, and Swaziland produce only negligible quantities of wheat.
[b] Confirmed by questionnaire.
[c] No data available.

Source: Anonymous, 1991b, and questionnaire

lation growth on the continent. Cereal yields per acre are generally the lowest in the world (FAO, 1990; Esterhuysen, 1992). Some countries have been particularly hard hit *i.e.* Zambia and Zimbabwe. Other countries already suffering food shortages due to war, *i.e.* Mozambique and Angola, now face famine. This has forced many countries to reverse attempts at restricting wheat imports placing further strain on their limited foreign exchange.

Anyone trying to estimate the food produced and the food available for human consumption in the African countries will face formidable problems. Few countries have land records, and the ones that exist only cover the most densely populated regions where cash crops dominate. In other parts of these countries, land is usually owned communally and production is mostly subsistence. The harvested area also tends to vary from year to year in response to rainfall and other vagaries of nature...Yields are also more difficult to estimate in Africa than in most other places... With practically no base data on acreage and yields derived from modern, scientific methods, it is of course, impossible to produce reliable estimates of staple food production (ADPRG, 1990).

Statistics relating to wheat production in Southern and Central Africa are, as stated earlier, difficult to come by and

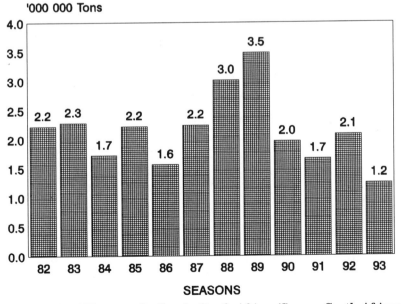

Figure 10-1. Wheat production in South Africa. (Source: South African Wheat Board, 1993)

often unreliable. The available information is shown in Table 10-1 and Figure 10-1, and to place these numbers in a more global context, South Africa, by far the largest producer in Sub-Saharan Africa, was ranked 30th on the world production list (PC Globe, 1992). The same source noted that Africa's largest producers were Morocco with 3.93 million tons (24th) and Egypt with 3.15 million tons (28th). Wheat import data is even harder to obtain but that which is available is shown in Table 10-2. A general overview of the situation in southern and central Africa is shown in Table 10-3.

Wheat Consumption, Import, and Production

Angola. The demand for wheaten products is high, particularly in the form of bread. Visitors to Angola will note that bread, hot from the oven, is sold out within minutes at what appears to be very high prices (Alex Jack, personal communication to the Wheat Board). France is the main supplier of wheat with the United States (US) attempting to compete. Flour imports are mainly from the European Community (EC). The lack of port facilities is, however, a major constraint. The State owns the majority of the mills and bakeries though privatization is believed to be imminent. The Government is also encouraging participation by South African commercial bakeries. The Angolan civil war has cost the country dearly, resulting in poor realization of its potential as a wheat producing country, and food basket for Africa.

Table 10-2. Wheat and Flour Imports for Southern and Central Africa (1,000 t), 1989–1993 (Projected)[a]

Country	1989	1990	1991	1992	1993
Angola	250	150	140	140	140
Botswana	...[b]	...	95	115	115
Lesotho[c]	48	50	53	56	56
Namibia[c]	50	50	50	50	50
South Africa[c]	0	494	552	257	1,000
Swaziland[c]	30	30	30	30	30
Zaire	270	270	270	270	150
Zimbabwe	50	50	50	50	355

[a]Kenya, Mozambique, and Zambia have only average data available, which is discussed in the text. No information is available for Malawi and Tanzania. Nigeria, officially, imports no wheat.
[b]No data available.
[c]Confirmed by questionnaire.

Source: U.S. Wheat Associates, 1992

Table 10-3. General Overview of the Economic Situation in Southern and Central Africa

	Angola	Bot-swana	Kenya	Le-sotho	Malawi	Mozam-bique	Nam-ibia	Nigeria	South Africa	Swazi-land	Tan-zania	Zaire	Zam-bia	Zim-babwe
GNP ($ U.S. million) in 1991[a]	7,160	1,446	9,484	847	1,568	1,077	1,564	27,751	88,629	746	3,216	9,001	2,969	6,384
GNP growth rate[b]	9.2	10.2	3.9	1.7	3.1	−5.0	2.9	−1.0	1.5	4.5	2.2	0.9	−1.5	2.5
Area (million km²)[a]	1.247	0.600	0.583	0.030	0.118	0.802	0.824	0.924	1.221	0.017	0.945	2.345	0.753	0.391
Population in 2000 (mil)[b]	13	1.5	36.6	2.2	12.3	21.7	1.8	157	47.8	1.1	37	48.4	11.3	13
Population growth rate (%)[b]	3.1	2.2	3.8	2.4	3.5	3.0	2.8	3.0	2.3	3.1	3.3	3.0	3.4	2.5
Population density (per km²)[a]	7	2	43	59	80	19	2	96	33	49	28	16	11	27
Degree of urbanization (%)[a]	26	24	22	19	15	23	27	16	58	26	20	40	47	26
Official language(s)[a]	Portg.	English	Swahili	Sotho (S.), English	Chewa, English	Portg.	Afr., English	English	Afr., English	Swazi, English	Swahili, English	French English	English	English

Botswana. Bread is a staple food and Botswana imports all its wheat requirements. Traditionally Botswana imported from South Africa and EC but now US wheat is also entering the market. Botswana has two major mills operating as Bolux Flour Mill Pty Ltd. and Francistown Milling Company Ltd. The latter is linked to those in Mauritius, Swaziland and Namibia through common ownership by Namib Management Services which has also commenced operations in Mozambique (7.5 t/h) in May 1993. Namib Management Services, using trade facilities such as General Sales Manager (GSM) and Export Enhancement Programme (EEP) is now able to bulk purchase and distribute to all its mills in Southern Africa, with a combined operating capacity of 260,000 t. Botswana will be increasingly important in this process.

Kenya. Though 50% self-sufficient in wheat, grown as a commercial crop by large-scale farmers in the Rift Valley province and Norak district, demand is increasing rapidly. This increased demand is due to the high population growth and increased urbanization. Kenya is emerging as the leading East African wheat consuming nation with a per capita consumption of approximately 22 kg per year which has increased 50% over the last ten years (Streak and Richter, 1992). Kenya imports approximately 300,000 t per annum from Saudi Arabia, USA, EC and Australia. Major millers include Nairobi Flour Mills Ltd. and Nakuru Flour Mills Ltd. Major bakeries include Cookies Ltd., Elliots Bakeries Ltd. and House of Manji.

Lesotho. Though bread consumption is rising steadily in the mountainous kingdom of Lesotho it has very little capability to produce wheat. Imports are mainly from Saudi Arabia, Argentina, EC and occasionally South Africa. As Lesotho is now eligible for EEP the US is expected to join this market. Imports are via the South African infrastructure (ports and rail links). The major bakery is Lesotho Bakery Pty Ltd. Lesotho Flour Mills, established in 1979, supplies all the nation's flour and sugar and half the maize. The mill is run by the UK firm Spillers Milling Ltd. A second mill, near Ficksburg, with South African links, has commenced production in mid 1993.

Malawi. Malawi has negligible wheat production but demand for bread is increasing. It imports wheat from Canada, South Africa, USA and EC. Malawi has two mills (total capacity < 6 t/h) operating in Limbe. As the baking industry has a requirement of approximately 3,000 tons/month some flour imports are also required.

Mozambique. Though bread is the traditional staple, the

war has ensured there is no production, nor foreign exchange, for imports. Most of the wheat available is in the form of food aid from Canada. Mozambique's estimated market is for 500,000 t per annum, though the estimated wheat import capability for 1992 is only 125,000 t. Almost all of the milling and bakery structure is centered on Maputo (4 mills and 36 bakeries) and in mid 1993 only 10% of the milling capacity was being utilized.

Namibia. Namibia imports 90% of wheat and flour requirements. Eligibility for EEP has moved imports away from South Africa to the US. EC and Saudi Arabia also compete in this market. Namib Management Services controls the only commercial milling operation in the country.

Nigeria. Demand remains high though wheat production meets only 5% of requirements. Nigeria has operated a wheat import ban since 1987 though certain business interests have managed to establish alternative cross border channels. Once boasting 22 flour mills, most have now closed down.

Swaziland. Swaziland faces the problem of negligible wheat production but a rising demand. Imports are expected to exceed the current figure of 40,000 t, mainly from France and the US, due to COFACE credits and EEP eligibility. The traditional supplier, South Africa, will probably lose its leading position as a result. The only commercial wheat flour mill, Ngwane Flour Mills Ltd., was opened in 1992.

Zaire. A massive 96% of demand is imported, mainly from the US. There is only one major milling operation (60 t/h), Minotorie de Matadi (MIDEMA) and, like the three largest bakeries (one of which is one of the world's largest), is sited in Kinshasa.

Zambia. Demand continues to outstrip supply. Average imports exceed 38,000 t/year, mainly from South Africa, Australia and US. The drought is expected to result in an greatly increased import requirement. All three mills are owned by the National Milling Company Ltd. and have a total capacity of 86,000 t/year. Major bakeries include Supa Baking Company Ltd. and Carousos Bakery Ltd.

Zimbabwe. Normally self-sufficient in wheat, though this situation may change as demand is steadily rising. The refusal of government to release foreign exchange has resulted in this demand going unmet. Drought has recently forced importation to be considered (approx. 350,000 t), with Canada, EC and US expected to be the main suppliers. Zimbabwe boasts 4 commercial wheat flour mills producing nearly all of the local requirement. The major milling companies are National Foods Ltd. (2 mills and 70% of the market), Blue Ribbon Foods Ltd. (20% of the market) and Midlands Milling Company Pvt. Ltd. (10% of the market). The baking industry was deregulated early in 1993, resulting in increased bread prices and a subsequent drop in consumption (G. Lutz, personal communication 1993). The remaining demand for bread plus limited wheat production requires a system of flour rationing and of blending maize and sorghum flour to wheat flour. Millers are required to purchase all wheat from the Grain Marketing Board and permission to import wheat is conditional on securing a flour export order, which can not be complied with due to insufficient local supply. Imports to Zimbabwe are via Durban (South Africa) and Biera (Mozambique).

South Africa. The average size of the annual wheat crop of about 2.2 million metric tons has varied between 1.2 and 3.5 million metric tons over the past ten years and is largely determined by the climate and rainfall patterns in the summer rainfall wheat production areas of the Orange Free State and Transvaal provinces of South Africa. The average national yield of 1.52 tons per hectare includes high yields obtained under irrigation and in the Cape Province winter rainfall production areas. This reflects the often unfavorable conditions for wheat production in the summer rainfall areas. The breeding and release of wheat varieties adapted to the different conditions in the various wheat production areas has played a major role in ensuring the future of wheat production in South Africa.

The annual requirement for wheat for human consumption has remained constant for several years at approximately 2.2 million metric tons, despite the annual population growth rate of 2.3%. South Africa has exported small quantities of wheat to several African countries on a regular basis. However, due to the ongoing drought between 1989 and 1992 the South African Wheat Board has imported wheat mainly from Canada, US, Australia and the EC. Eligibility for EEP has moved the market in favor of US wheat.

Flour Milling. South Africa boasts six major flour milling groups, Premier, Tiger, SASKO, Ruto, Bokomo and Genfood, operating 34 modern flour mills with a milling capacity ranging from 3.5 to 42 t/h. These mills are responsible for the production of more than 98% of the flour used in South Africa, with the first three groups mentioned above accounting for 75%. Most of these mills are highly automated, despite the availability of relatively inexpensive labor. One of the ten largest flour mills in the world, Ruto Mills Pty. Ltd., with a production capacity of 42 t/h, is situated in Pretoria. About 20 small independent flour mills produce the remainder of the local flour requirement.

Although approximately 2.2 million tons of wheat are milled annually into flour, a relative small number of different flour types is produced. White bread flour (~78% extraction) and brown bread meal (~87% extraction), constitute 90% of flour produced, and are used mainly in the production of white and brown pan bread. Unlike other African countries, South Africa, and its neighbors Namibia, Lesotho and Swaziland, produce brown bread containing selected bran at levels ranging from 9 to 15% by mixing the bran with white bread flour to produce brown bread meal.

Although strong preference exists for white bread, financial factors have influenced the market. Brown bread prices have been, since the 1980s, usually abnormally lower than those for white bread due to either higher subsidies (prior to deregulation) or 0% VAT (post deregulation). These influences have distorted the market from a probable brown/white ratio of 20/80 to its current 46/54. Up to the 1989/90 season, wheat and bread consumption was increasing at a rate of 2 to 4% per annum but has since declined as a result of the economic and political environment. This has been exacerbated by the prevailing drought.

Other flour types include cake flour (ash content <0.55% and Kent-Jones [KJ] color grade <1 KJ units) and "all purpose" flour (color grade of <2.2 KJ units). These flours are low in ash content and exhibit a better Kent-Jones flour color grade in comparison to white bread flour (3.5 KJ units). Limited quantities of high ratio cake flour and self-raising flour are also manufactured.

Baking Industry. Traditionally, the baking and milling industry was regulated and the selling price of wheat, flour and bread was controlled. Government subsidy on the production of standard bread, together with a policy of rationalization of the milling and baking industries by means of compulsory registration and licensing played a significant role in the establishment of large milling and baking companies where 85% of bakeries were owned by the milling groups well equipped for the production of subsidized standard white and brown pan bread. Several plant bakeries in South Africa boast production capacities exceeding 8,000 loaves per hour.

Most of these large plant bakeries adopted the Chorleywood or no-time dough bread-making methods common in the UK since about the 1970s. Formula and processing are similar in most bakeries producing standard bread.

A typical formula for commercial pan bread (based on 100 kg flour weight) is given below:

Flour	100 kg
Compressed yeast	1.80 %
Salt	2.00 %
Sugar	0.50 %
Fat	0.40 %
Emulsifier	0.20 %
Soya flour, full fat, enzyme active	0.50 %
Preservative	0.15 %
Ascorbic acid	0.004%
Potassium bromate	0.002%
Fungal alpha-amylase	0.008%
Water	60.00 %

Regulations on the dry solids content and fat content of the so-called "government" or subsidized bread were strictly enforced. Subsidized bread was relatively cheap, uniform and of good quality.

With the deregulation of the milling and baking industries in March 1991, the government subsidy on bread, already being phased out over a number of years, was lifted and the nominal mass of white and brown pan bread was reduced from 850 g to 800 g. Large numbers of smaller bakeries, in-store bakeries and hot bread shops mushroomed in urban areas to take part in the deregulated market. Notwithstanding this development, the larger and better equipped plant bakeries retained their dominant role in bread production. The technical expertise, existing infrastructure and bread distribution capabilities aided plant bakeries in retaining approximately 75% of the pan bread market share.

Approximately 90% of all flour milled, is used in the production of white and brown pan bread. South Africa is also unique in Africa, and possibly the world, in that 46% of all bread consumed is in the form of brown pan bread.

Other forms of bread such as soft and crispy bread rolls often topped with caraway, poppy or sesame seed are also popular. The diversity of the South African population has brought with it demand for various wheaten products ranging from the typical French baguette, pita bread and pizzas from Italy to the *challa*, a plaited Jewish ritual loaf enriched with butter and eggs giving the creamy colored crumb and a very fine texture. Most bread types produced in South Africa originated in Europe. Breads such as *kitke, coburg, bloomer*, soda bread and a variety of rye breads are also available throughout South Africa.

Popular Products Made from Hard Wheat

Three generic types of bread (Table 10-4), using typical recipes as shown in Table 10-5, are produced in the African countries shown in Table 10-6.

Production Methods. Whilst Africa can boast one of the world's largest bakeries in Kinshasa it also has some of the most primitive. The bakery at Kinshasa is equipped with the latest in German baking technology, is fully automated and processes approximately 220 tons of flour per day. The most

Table 10-4. Bread Types Produced in Central and Southern Africa

British	Mainly pan breads (open top and sandwich), sugar breads, and soft breads
French	Baguettes and boulot all with French cutting style
Portuguese	Bolochas, basically French recipe but with different finishing and cutting style

Table 10-5. Bread (Generic) Recipes Used in Central and South Africa (Based on 100 kg Flour Weight)

	British	French	Portuguese
Flour (kg)[a]	100	100	100
Salt (%)[b]	1–2	1–2	1–2
Yeast (%)[c]	>1	0.3–1	0.3
Water (%)[d]	±60	±55–57	±60
Improver (%)[e]	0–1	0–1	0–1
Fat (%)[f]	0.5–6	...	0.5
Sugar (%)[g]	0.5–12	...	...

[a] Only white bread flour used.
[b] A valuable ingredient in Africa, which is relatively expensive and less readily available as a result. Usage is very dependent on price, and as a result, lower usage levels are more common. Coarse salt is often a problem, requiring dissolving before use to avoid dough collapse.
[c] Instant yeast is most widely used, again very price dependent.
[d] Water quality varies greatly in pH and hardness.
[e] Improvers are used depending on availability, the prevailing environmental conditions, and price. Very few producers use improvers in the British-style recipe, except in South Africa where the use of improvers is common.
[f] Shortenings are used wherever possible; if not available, oils are used.
[g] Very price dependent. Usage is higher in countries where sugar is produced.

primitive consist of a wood heated clay oven in which the dough pieces are stuck to the wall for 10 minutes.

Throughout Africa one of the major problems faced by the baking industry is the maintenance of equipment. In almost all cases (Kinshasa included) this is carried out by bakery personnel using whatever tools and spare parts they can obtain. The lack of foreign exchange for spares is only part of the problem; most of the equipment is so old that spare parts are no longer available. Even if they were, there are countless difficulties still to be overcome. It can take months to obtain those spares.

Mixers. Throughout Africa the most common type of mixer in use is the French Mixer estimated (M. Vael, personal communication) at 80%, with the Artofex and Diosna mixers making up most of the remainder. The French Mixer typifies the problems faced by bakery personnel. The original models operated at two speeds (slow and fast), the African version invariably has only one speed still working. The rotating mixing bowl usually has a broken clutch so that the bowl now freely rotates but this problem is usually solved with typical ingenuity by resting a bag of flour against the bowl to act as a brake. Mixing times are typically around 20 minutes on the slow speed setting and 12 - 15 minutes at the higher speed.

Prefermentation. The French style recipe requires no prefermentation whereas the Portuguese style requires long prefermentation, typically 2 hrs, due to the low yeast levels and

Table 10-6. Classification of Bread Styles Produced in Africa

Belgian	British	French	Portuguese
Burundi	Botswana	Algeria	Angola
Rwanda	Egypt	Cameroon	Guinea Bissao
Zaire	Gambia	Central African Republic	Equatorial Guinea
	Ghana	Chad	Mozambique
	Kenya	Comores	Sao Tome
	Lesotho	Congo	
	Liberia	Gabon	
	Malawi	Gambia	
	Nigeria	Guinea	
	Seychelles	Ivory Coast	
	Sierra Leone	Madagascar	
	Somalia	Mali	
	South Africa	Mauritania	
	Tanzania	Mauritius	
	Togo	Morocco	
	Uganda	Niger	
	Zambia	Reunion	
	Zimbabwe	Senegal	
		Sierra Leone	
		Tunisia	

recipe formulations. The British style recipe depends on the technique used varying from the sponge dough process (with prefermentation times as long as 5 hrs) in the bigger bakeries to bulk fermentation in the smaller ones.

Provers. It is estimated (M. Vael, personal communication) that 80% of bakeries do not have provers (proof cabinets). These bakeries utilize the heat from the sun, covering the dough pieces with flour bags or whatever is available. Ambient conditions in Africa are such that some of the bakeries possessing provers do not switch them on but instead place the dough pieces in the cabinet as it stands.

Ovens. It is a generality, but nevertheless essentially true, that the type of oven available dictates the type of recipe used. British style recipes are only produced in the major centers and utilize Deck and Rotary ovens, with the big bakeries using traveling ovens (usually only one per city, but not found in every country). French style recipes are baked in Deck and Rotary ovens in 90% (M. Vael, personal communication) of cases with stone ovens making up the remainder. Portuguese style recipes are baked, for approximately 20 minutes, in stone ovens.

Fuel for these ovens depends on availability. Electricity is only used where readily available, normally the bigger cities only. Gas is non-existent. The most common fuel sources being diesel, for Deck and Rotary ovens, and wood for stone ovens.

Quality. Consumers are more quality conscious than expected to the extent that they will be reluctant to purchase bread if it does not match their perception of quality. To the African the most important quality criteria are product color and volume, perceptions many readers, especially those with no first hand experience of Africa, will have difficulty coming to terms with. Consumer resistance to change is strong. If the product is not as they expect they will not buy it and bakery owners (mainly Europeans, especially Greek in Central Africa) have long recognized this.

Distribution. Very few countries in Africa have the type of distribution system found in Europe, North America and South Africa. The majority of countries rely on hawkers and street vendors. These people will come to the bakery and purchase bread (quantity dependent on their financial position) which they will then sell on street corners or door-to-door at a small profit. Having sold their bread the process will be repeated as often as possible as this is, usually, their only source of income. In some of the larger centers, and for some of the outlying areas, the bakery will deliver to a central distribution point at which point the hawkers and street vendors take over.

Flat breads are to be found, more usually, in the rural community. The recipe used is based on the French style with no improver, very little yeast (typically 0.1%) and, if available, 0.5% oil.

Very little, if any, composite flours are used in the breadmaking process within the urban communities with the exception of Zimbabwe. In Zimbabwe, as already stated, quantities of maize and sorghum flour are routinely combined with wheat flour, the ratio depending on availability of wheat.

Popular Products
Made from Soft Wheats

Traditional products of soft wheat are virtually non-existent in Africa except for South Africa and Namibia. Breakfast cereals, pies, pastries, soup thickeners *etc.* can be found but are generally imported and very expensive. Some specialty shops, in the major cities, do exist but they cater more to the tourist and visitors than to the local population. Soft wheat, as such, does not exist in Africa and the other ingredients, such as fat and sugar, are not always readily available.

Notwithstanding the above, sponge cake and cookie biscuits are popular and widely available. Sponge cakes are traditionally made in large flat trays, flavored in accordance with local availability of flavors, often in a marbled format. These trays are then cut into slabs for retailing. The traditional British type of wedding cake may also be found on such occasions but it is more usually a sponge cake covered with multi-colored icing. Cookies are found in a very simple format and are usually of the crumbly or hard type. In the major cities some cookies may be coated on one side with chocolate.

Popular Products
Made from Durum Wheat

Although very little durum wheat is produced in Africa, South Africa has produced up to 29,000 t per annum. Lately, due to high producer costs, most of the local durum requirement is imported. The pasta industry in South Africa is dominated by Fatti's and Moni's, part of the Tiger Milling group. In addition several companies in Botswana also produce pasta products for both their local and South African markets. In several other countries, those with available foreign exchange, pasta products are imported.

Consumer Trends

The consumption of bread in Africa is increasing and, due to the demands of the local populations, so are wheat imports. Part of this increase appears to be due to the price differential between locally produced, and traditional, carbohydrate sources, such as maize and rice, and imported wheat (Dendy, 1992). The other reason for the increase is availability. Once the staple of only the elite bread has now become the staple of the masses. The demand for bread has led to many hundreds of small bakeries being established which in turn enabled more people to sample, and to experience, the delights of bread. The wheel has continued turning and more bakeries are being established to feed that demand. The only constraint appears to be the availability, and price, of wheat.

Acknowledgments

The authors gratefully acknowledge the personal experiences of many friends and colleagues who have traveled through Africa which in the retelling enabled us to confirm some, otherwise, undocumented material. We are particularly grateful to Marc Vael, Ceres Soufflet Group, Belgium, who gave up so much of his free time to recount his experiences in Africa over the past decade. Special thanks are also due to Alex Jack and Johan Potgieter for describing their experiences on recent visits within Africa. The authors also acknowledge Judith Streak and Nici Richter, INFOTEK, CSIR, who produced a general overview of Southern Africa, under a great deal of time pressure, which gave us a flying start.

References

ADPRG. 1990. African Development Perspectives Yearbook. African Development Perspectives Research Group. Pub. Schelzky & Jeep, Berlin.

ANON. 1990. The story of wheat in South Africa. Pub. Wheat Board, Pretoria.

ANON. 1991a. World food supplies. Background Brief, January.

ANON. 1991b. Country surveys: Angola, Botswana, Mozambique, Tanzania, Zambia, Kenya, Malawi, Zimbabwe. Africa Today.

Dendy, D.A.V., 1992. Perspectives in composite and alternative flour products, in cereal chemistry and technology: A lond past and a bright future. Proc. ICC 9th Cereal and bread Congress, Paris.

Esterhuysen, P. 1992. Africa at a glance: Facts and figures. Africa Institute of South Africa, Pretoria.

FAO. 1990. Food and Agricultural Organisation Yearbook: Production 1989. Rome.

FAO. 1992. Food and Agricultural Organisation Yearbook: The state of food and agriculture 1991. Rome.

PC Globe. 1992. PC Globe Version 5.0, PC Globe Inc., Tempe.

Sanai, D. 1992. SA and Zambia join forces. Business Day, 20th March.

Streak, J. and Richter, N. 1992. Wheat, its end users and uses in Southern Africa, Kenya, Zaire and Nigeria. Survey commissioned by the Wheat Board, carried out by INFOTEK, CSIR, Pretoria.

US Wheat Associates. 1992. Marketing Plan FY 1993, Subsaharan Africa. Pub. US Wheat Associates, Abidjan.

World Bank. 1990. The World Bank Atlas, Washington DC.

CHAPTER 11

Wheat Usage in Australia and New Zealand

Graham J. McMaster and John T. Gould

WHEAT USAGE IN AUSTRALIA

Introduction

Wheat was first cultivated in Australia in New South Wales at the time the country was first settled by the English in 1788 (Macindoe and Walkden Brown 1968). Tall growing and susceptible to many diseases, the cultivars that were sown were not adapted to Australian conditions. A large agricultural industry has arisen from this humble beginning. Concerted plant breeding efforts have resulted in modern day cultivars well adapted to the dry Australian climate and possessing good disease resistance, agronomic qualities and grain quality. Australia is now the third largest wheat exporter behind the USA and Canada (excluding the European Economic Community). In 1988/89 and 1989/90 wheat constituted 5% of the value of total Australian exports and contributed over $A2 billion to the national economy.

Wheat Production

As detailed in a recent international survey (Crofts 1989), definitions of spring and winter wheats vary between countries. The normal growing period in Australia is shown diagrammatically in Figure 11-1. In Australia, wheat is a winter crop and is generally sown between May and July. Because of winter temperatures, Australian grain growers sow "spring types" of cultivars. Only a relatively small amount of "winter types" are grown and these are sown in the earlier months of March and April.

The majority of the wheat belt lies in a zone which receives 230–380 mm of rainfall during the normal growing season.

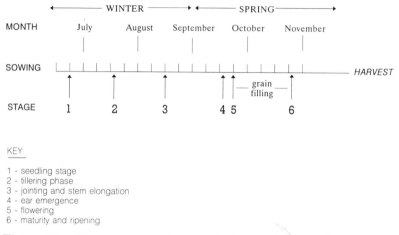

Figure 11-1. Diagram of growing period of wheat in Australia.

Figure 11-2. Wheat growing areas in Australia.

Different classes of wheat are grown across Australia as shown in Figure 11-2.

The size of the Australian wheat crop is approximately 15 million tons. The average production figures for each State over a 10 year period are given in Table 11-1.

The two largest wheat producing States are Western Australia and New South Wales. The large variation in production figures (i.e. maximum and minimum production) over the 10 year period illustrates the seasonal difficulties facing Australian grain growers. Rainfall is highly variable in Australia and to ensure a good yielding crop, rains are required for sowing and then at appropriate intervals throughout the season. At the time of harvesting, October through December, (depending on the region) day length is increasing and temperatures rising with the onset of summer. More often than not, this places crops under moisture stress. The average yield of wheat crops in Australia is shown in Table 11-2.

The average Australian wheat yield of 1.41 tons per hectare is quite low compared to the high yield environment of Europe and reflects the drier, hotter summer conditions experienced in the majority of the Australian wheat belt.

Table 11-1. Annual Production of Wheat (1,000 tons) by State (10-Season Period, 1981–1982 to 1990–1991)

	Average	Minimum	Maximum
New South Wales	4,819	1,476	8,819
Victoria	2,123	387	3,908
South Australia	1,875	681	2,798
Western Australia	4,949	3,820	6,476
Queensland	1,372	706	1,972
Tasmania	3	1	5
Australia	15,144	8,736	21,669

Source: Australian Bureau of Statistics

Table 11-2. Wheat Yields (tons/hectare) by State in Australia (10-Season Period, 1981–1982 to 1990–1991)

	Minimum	Maximum	Average
New South Wales	0.46	2.20	1.58
Victoria	0.29	2.42	1.71
South Australia	0.48	1.79	1.25
Western Australia	0.89	1.55	1.22
Queensland	0.96	1.98	1.55
Tasmania	1.63	5.34	2.83
Australia	0.75	1.67	1.41

Source: Australian Bureau of Statistics

Wheat Marketing

The Australian Wheat Board (AWB), under the powers of the Australian Wheat Marketing Act of 1989, has the sole responsibility for the export marketing of Australian wheat. Australian wheat is broadly classified into five grades: Prime Hard, Hard, Standard White, Soft, and Durum. Classification standards based on quality characteristics are set by the AWB each year. The AWB may further separate these grades into a large number of classes according to cultivar, protein content and other quality characteristics for marketing purposes and pay wheat growers on that basis. Whilst Australian wheat is exported in a regulated environment, the domestic market is a deregulated environment. Wheat growers have the choice of selling their wheat to grain traders, directly to flour mills, or to the AWB.

End Uses of Wheat

Export. Approximately 85% of the wheat produced in Australia is exported by the AWB to over 50 countries where it is milled and processed into a diverse range of end-products. The major product of the Middle Eastern markets is the range of Arabic flat breads. The major products made from Australian wheat in the Asian markets are the various types of noodles: Chinese (yellow alkaline) noodles, Japanese (white salted) noodles and instant noodles. Chinese steamed breads and dumplings are also made from Australian wheat in China and other parts of Asia.

Whilst it is very difficult to estimate the relative proportion and/or importance of these products as markets and volumes change from year to year, estimates have been attempted by the Wheat Science and Technology Division at the Bread Research Institute of Australia (BRI). An estimate of the relative proportion of Australian wheat used in the manufacture of various products in international markets is given in Table 11-3.

Since these estimates were made, the Middle Eastern markets have become more important with up to 50% of wheat exported to these markets prior to the Gulf war. More recently, the specialty markets of South East Asia, North Asia and China have received a great deal of attention, reflecting the increased Westernization in food consumption patterns.

Knowledge gained from research has given a much better understanding of the qualities and nature of the various types

Table 11-3. Uses of Wheat (%) Exported from Australia[a]

Arabic bread types	32
Alkaline (Chinese) noodles	22
White (Japanese) noodles	11
Pan bread	21
Steamed bread	4
Cakes and biscuits	2
Feed	8

[a] Ten-year average to 1986; estimated on a total wheat crop of 14 million tons.

Source: Miskelly, 1988

of noodles and flat breads. Research has also led to the development of small scale tests and the ability to manufacture products at a test (small scale) level. Products have to be assessed and ranked according to quality: hence it was necessary to establish scoring systems and sensory evaluation panels. The primary focus of the research has been to:

- Establish the important quality characteristics for each product.
- Develop methods to manufacture these products at a test (small scale) level.
- Develop sensory evaluation and scoring methods to rank products for overall quality characteristics.
- Establish the appropriate flour quality characteristics for specific end-products.
- Develop appropriate (optimal) wheat quality parameters for specific end-products.

Research involving Asian noodles (Crosbie et al 1990), Arabic flat breads (Quail et al 1991a,b) and Chinese steamed breads (Huang and Miskelly 1991) has played a major role in the marketing of Australian wheats. It has led to increased product differentiation (specific classes and grades based on processing performance in niche markets) and has increased the capacity to tailor quality of wheats to meet customers' needs in the international marketplace.

Domestic. The majority of the wheat utilized in Australia is used for the production of flour for human consumption or for the production of starch and gluten. A relatively small amount is used for animal feedstuff, although in times of drought when availability of other grains is limited, wheat is often used.

Approximately 1.8 million tons of wheat is processed in

Australian flour mills annually. In 1990 1.6 million tons of flour was produced of which 70,866 tons was exported. The Australian market is generally regarded as a "mature" market. However, flour production has increased by approximately 196,850 tons over the five year period 1986–1990. Trends in the marketplace are given in Figure 11-3.

Australia is a major manufacturer of gluten, producing 47,918 tons in 1989/90 and exporting 26,627 tons in the same period. Other industrial use of flour in Australia is only minor. Of wheat flour that is used for purposes of human consumption, the major end-use is for bread products. The consumption of bread in Australia has been steadily declining over the past decade. However, it now appears to have stabilized at approximately 45 kg per capita per annum. The other major end-uses for wheat flour are for pastry goods, biscuits, pasta, packeted flour and mixes, and general use in the processed food industry. The end uses of flour in Australia are given in Table 11-4.

The total value of the baked products industry in Australia is approximately $A2.86 billion. This figure, however, does not include fresh pies which are a very popular pastry product in

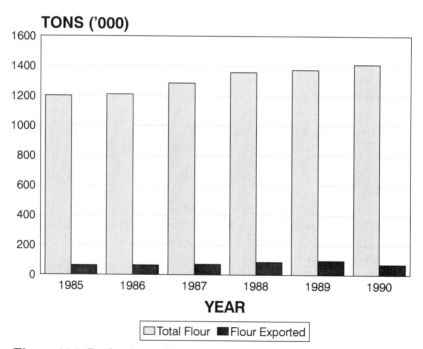

Figure 11-3. Production and export of flour—trends in the marketplace.

Australia. Estimates of the economic value of product groups based on retail sales is given in Table 11-5.

Bread Products. Bread products, with a retail value of approximately $A1.6 billion, are the most economically significant baked products in Australia. The range and variety of bread products made in Australia has increased dramatically over the past few decades. This is partly a reflection of the migration patterns that have taken place to form the current multicultural society in Australia and the subsequent establishment of ethnic bakeries.

Whilst the range of baked products has increased, white sliced bread remains the dominant segment of the market. There has been a trend in more recent years towards "high-fibre" and "multigrain" breads in line with the positioning of bread as a component of the healthy diet.

Table 11-4. Flour Usage in Australia

	1989		1990	
	Tons	%	Tons	%
Industrial				
Starch/gluten manufacturers	303,146	24.0	341,601	26.0
Other industrial users	1,504	0.1	1,684	0.1
Human consumption				
Bread	586,269	46.6	585,755	44.6
Pastry	70,642	5.6	71,674	5.4
Biscuit	76,658	6.1	79,214	6.0
Pasta	39,196	3.1	52,522	4.0
Packeted flour and mixes	80,989	6.4	77,322	5.9
Food manufacturers	102,027	8.1	105,610	8.0
Total	1,260,431	100.0	1,315,402	100.0
Export	91,189		70,993	
Grand total	1,351,620		1,386,395	

Source: Bread Research Institute of Australia Inc.

Table 11-5. 1990 Retail Grocery Sales in Australia[a]

Product Category	Value $A M
Bread	1,630
Cakes (fresh)	187
Biscuits	675
Cake flour/mixes	78
Frozen bakery products[b]	288

[a] Fresh pies not included.
[b] Includes savoury pies, frozen desserts (fruit pie, sweet pastries, cold desserts, cakes, doughnuts/muffins), pizzas, unbaked pastry, beefburgers/hamburgers.

Source: Retail World, December 1990

Other bread types include:

- brown and wholemeal breads
- protein increased breads
- milk breads
- Vienna and "continental" breads (baked on the oven hearth)
- fibre increased breads
- Arabic flat breads
- special diet breads
- Indian breads
- variety of bread roll products.

Breadmaking Process. The rapid-dough (no-time dough) system is used mainly in commercial bakeries in Australia (Australian Handbook of Breadmaking, 1989). In recent times, some new commercial bakeries have installed high intensity "Tweedy" mixers. Smaller bakeries tend to employ medium intensity "spiral" mixers. Low intensity mixers are not routinely used in bakeries in Australia.

The addition of suitable flour "improvers" is a feature of the rapid-dough system, which enables doughs to be mixed to full development quickly without the high mixing intensities needed in mechanical dough development (MDD) systems. The rapid-dough system, like the Chorleywood process, does not include a significant bulk fermentation step and bread can be manufactured in a total of 120 min.

The major elements of the Australian rapid-dough system of breadbaking are:

- rapid maturing agents (oxidants) must be included in the dough
- addition of some gluten softener is beneficial with stronger types of flour (especially when mixed in low speed mixers)
- fat in solid or plastic form is included in the dough
- full dough development must be complete at the end of mixing
- yeast content is slightly higher than for normal bulk fermented doughs.

Some typical Australian bread formulations are given in Table 11-6.

The Australian bread industry is currently made of three segments: wholesale bakeries, hot bread shops and "in-store" bakeries.

Wholesale Bakeries. These bakeries are large, automated and account for approximately 65% of the total bread market. The companies who own large plant bakeries are vertically integrated with interests in flour milling, bakery ingredients production and related industries. The four major companies include Quality Bakers Australia (Goodman Fielder), Tip Top (George Weston), Sunicrust (Bunge) and Regal/Cobbity Farm (Defiance Mills).

Hot Bread Shops. Hot bread shops have proliferated in Australia through the 1980s and have made significant inroads into the market share of the larger plant bakeries. The aroma of freshly baked bread has proved to be an extremely powerful marketing strategy. In 1976 in New South Wales there were only 13 hot bread shops. In 1988 this number had increased to almost 400. In Victoria the number is estimated to be on the order of 600 hot bread shops. A more recent phenomenon has been the emergence of franchise chains of hot bread shops offering a wide range of bread and pastry products.

In-Store Bakeries. In-store bakeries have become a significant market force in Australia. The bakeries are located in supermarkets and are major drawing cards for customers seeking a fresh, hygienic image. Both major Australian supermarket chains (Coles and Woolworths/Safeway) now have a policy of including a bakery in each new supermarket store. There are now estimated to be 280–300 in-store bakeries in Australia.

The evolution of the in-store bakery has introduced the concept that bread and bread baking is a food industry process where once it was considered separately as a baking industry process. It has already placed bread in the position of other

Table 11-6. Typical Australian Bread Formulations

Ingredients (parts by weight)	White Sandwich Bread	Milk Bread	Vienna Bread	Brown Bread	Wholemeal Bread
Flour	100	100	100	50	...
Wholemeal	...	...	...	50	100
Water (approx.)	58	60	59	60	62
Yeast (compressed)	2.5	2.7	2.7	2.5	2.5
Salt	2.0	2.0	2.0	2.0	2.0
Fat	2.0	2.0	2.0	2.0	2.0
Milk powder	4.0[a]	4.0	2.0	2.0[a]	2.0[a]
Sugar	1.0[a]	1.0[a]	2.0[a]	1.0[a]	1.0[a]
Dry gluten	...	...	...	1.0	4.0
Composite bread improver	---------------- as required ----------------				

[a] Optional.

Source: Bread Research Institute of Australia Inc.

grocery items in being a "loss-leader" in marketing terms when such a strategy is deemed appropriate by supermarket management.

WHEAT USAGE IN NEW ZEALAND

Introduction

In any country, the end uses of wheat will be the result of a complex inter-relationship of historical, social, cultural, geographic, agronomic, political, economic, technological and commercial factors. This is certainly the case in New Zealand. Since the commencement of European colonization in the early 1800s to the present, each of the above factors, at various times, has exerted a dominant influence. The result has been that wheat based food products of different type and quality have been produced at different stages of the country's history.

Wheat has been grown in New Zealand since the early 1800s with the first commercial flour mill being established in 1834. By far the great majority of early European immigrants to New Zealand were of British origin. As a consequence, until recently, most of the wheat flour products were of a type found in and made by methods commonly used in Great Britain. This situation was reinforced by the fact that many of the other factors listed above were transposed, with little modification, from Britain.

In recent years (and in common with many other countries) the range of products has become much more diversified. This has been the result of: immigration of people from countries other than Britain, an upsurge in international travel by New Zealanders, as well as the effect of a very food conscious news media. Since the mid 1980s, the speed of change has been dramatic with the range of bread types increasing almost day by day together with a significant increase in the use of pasta and other flour based products.

Wheat Production. Wheat growing in New Zealand is largely confined to the eastern (leeward) side of the South Island, which in 1990 provided 34,780 hectares or 85% of the 40,650 hectares sown in wheat in 1990. The actual distribution is shown in Figure 11-4. Of that area, 25,000 hectares or 61.5% was devoted to bread cultivars with the remainder being of cultivars being suitable for biscuits, pasta or stockfeed (New Zealand Dept. of Statistics Agricultural Census 1990).

Because of variable weather conditions in the growing areas, sprout resistance is of particular importance to New Zealand wheat farmers. The genetic make up imparting this characteristic is also associated with red wheat coloration. It is for these reasons that most cultivars grown in New Zealand are red wheat. This is in direct contrast to Australia where only white wheat may be grown. The bread wheats can be classed as semi-hard in endosperm texture.

An exception to the above is what is known as purple wheat. This wheat was originally grown for use in a wholemeal produced by a small Christchurch mill in the 1960s. Since then purple wheat, usually kibbled, has come to be widely used by

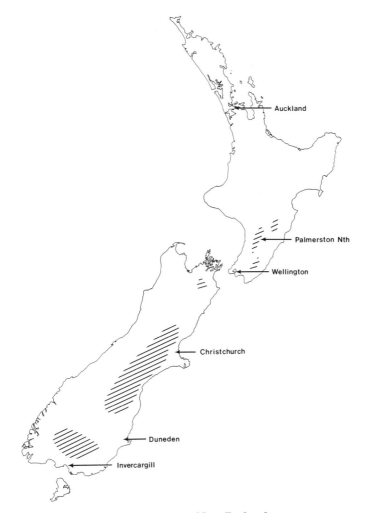

Figure 11-4. Wheat growing areas in New Zealand.

bakers in New Zealand to give a distinctive color effect to various specialty grain breads.

Since the early 1980s by far the greatest proportion (60%) of wheat is spring sown (planted August–September, harvested January–February). An interesting situation existed during the period 1988–91 when, because of the unavailability of suitable winter cultivars, the spring cultivar Otane was sown as a winter wheat with acceptable results. New winter wheat cultivars are now becoming available (DSIR Crop Profiles 1992).

Yields are high by world standards with the average overall areas in most seasons being in excess of 4 tons per hectare (DSIR Crop Profiles 1992).

The total domestic demand for wheat (393,700 tons) is minuscule compared with total world usage. Domestic production is even less (Table 11-7). It should be noted that since the deregulation of the wheat growing industry in 1987, the collection of statistics has been on an ad hoc basis. Figures for the period since 1987 may not be directly comparable with those prior to that date though significant trends will be obvious. The main use of wheat is for flour and related products with 295,275 tons being used by the milling industry (DSIR Crop Profiles 1992).

During the period 1936–87, wheat growing and utilization was carried out under the dominating influence of the New Zealand Wheat Committee and its successor the New Zealand Wheat Board. One of the principal objectives of those Government-controlled organizations was that New Zealand should be self-sufficient in wheat. Despite this objective, during control by the Committee and Board, there were only three years, 1968, 1972, and 1975, when New Zealand was self sufficient in wheat (New Zealand Wheat Review 1979). It should also be

Table 11-7. 1980's New Zealand National Areas, Yields, and Production Values

	Area (1,000 hectares)	Yield (ton/hectare)	Production (1,000 tons)
1979–1981	85	3.593	304
1982	72	4.019	287
1983	74	4.310	318
1984	64	4.528	289
1985	72	4.241	305
1986	92	4.084	374
1987	83	3.996	331
1988	51	4.007	202
1989	38	3.509	132
1990	50	4.330	216

Source: DSIR Crop Profiles

noted that this was a period when quality was not the critical criterion as it is today.

In New Zealand, wheat growing competes for land use with other arable crops as well as pastoral farming. Hence, the comparative dollar return for each of those can significantly affect the amount of wheat grown. There are no New Zealand farmers who rely on wheat for more than 50% of their income.

In more recent times (since the mid 1980s) the bias of breeding emphasis has been reversed. It is to the credit of the wheat breeders of the Crop & Food Research Institute (formerly the Crop Research Division of the Department of Scientific and Industrial Research, DSIR) as well as private breeders that they have been able to breed cultivars which have significantly improved bread making quality while maintaining yield (Figure 11-5). Equivalent bake scores for North American wheats are Hard Spring 26–28 and Hard Winter 24–26 (G. Tempest, Private Communication).

Until 1986 the general bread baking quality of available cultivars was marginal. As a result, small variations in growing conditions could have a disproportionate influence on end use quality. The release of the DSIR cultivar, Otane, in 1986/87 together with subsequent cultivars of similar bread baking qual-

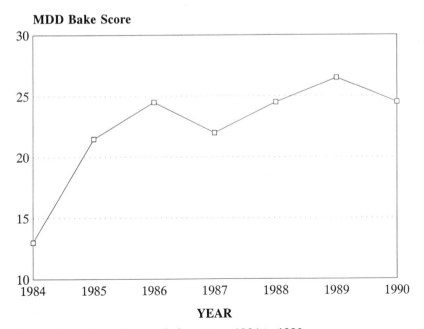

Figure 11-5. Mean harvest bake score—1984 to 1990.

ity, have raised the overall quality level and, consequently, improved the reliability for bread baking.

There is now a greater emphasis on quality parameters by all wheat and flour users (not just bread bakers). Now that quality parameters can be more clearly defined, there is likely to be more wheat production for uses other than standard milling grade.

The price of wheat varies with the world price, the most significant benchmark now being the price for Australian Standard White (ASW). The exchange rate also has a significant effect on price.

As the market for New Zealand grown wheat is, for all practical purposes, entirely domestic and completely deregulated, there is no need for an overall system of grading. There is a direct relationship between miller and grower and for milling wheat, mills contract with growers using a quality index system which allows a premium for top grade wheat. The price calculation takes into account all or some (depending on the contracting mill) of the following criteria: cultivar, bake score, protein content, sprout index, kernel weight, moisture content, screenings, location and delivery time.

End Uses of Wheat

Wholesale Bakeries. Bread flour is by far the largest proportion (57%) of flour milled in New Zealand. Wholesale bakeries, the predominant producers of bread (approximately 80% of production), mostly utilize a Mechanical Dough Development (MDD) method based on the Chorleywood Bread Process (CBP). This was first introduced to New Zealand by the author (J Gould) in 1964 and after an initial period of adaptation and refinement, rapidly gained acceptance. The principal reasons for the ready conversion from the previous (1–4 hour) bulk fermentation were as follows. First, New Zealand wholesale bread characteristics were, and are, very similar to those in Britain where the CBP originated. Second, flour quality for bread baking in the 1960s, 1970s and early 1980s was by world standards very poor. The CBP was developed in Britain to utilize the lower quality (cf. North American) wheat produced in that country.

For these two reasons the CBP was almost tailor made for New Zealand conditions. One major difference was that flour from New Zealand wheats required a much lower, and more variable, work input than did British flours (down to 5 wh/kg compared with 11 wh/kg in Britain). This required modification of the original mixer impact plates received from Britain. It also required the variation of work input to the dough by bakeries to take into account the variations experienced in different shipments of flour.

Recent (post 1987) improvements in the breadmaking quality of flour from New Zealand wheats have been accompanied by an increase in mixing requirements (work input). Optimum work inputs of up to 16 wh/hr. per kg are not uncommon. This has become critical for some bakeries. Consequently, wheat breeders are being requested to produce cultivars which have lower work input (12 wh/kg) requirements without sacrificing bread making quality. Equivalent work input measurements for North American wheats are hard spring 16–20 wh/kg, hard winter 14–18 wh/kg (G. Tempest, Personal Communication).

Bread made by wholesale bakeries commonly has a specific volume in the range 4.5–5 ml/g for the most popular lidded (Pullman) types of bread. Some less common unlidded varieties may have a specified volume of 6–6.5 ml/g. Wholesale bakeries produce a wide range of breads. The increasing emphasis on the healthiness of bread by nutritionists, supported by significant marketing budgets of bakeries, has resulted in between 50 and 55% of plant bread containing meals or grains of one type or another. These breads require gluten fortification to enable them to attain the specific volumes referred to above. With one or two very minor exceptions, all those breads are bagged in polythene bags.

The wholesale bread market is largely influenced by the activities of the three major supermarket chains in New Zealand. This influence is exerted via their policies of using the (low) price of bread to attract customers together with their bargaining power with regards to private label bread. As there are also three main wholesale bakery groups, each keen to maintain market share in what is a small market, trading becomes volatile at times.

The bread market is completely deregulated. There are no statutory restrictions on when bread may be made, sold, distributed, or priced. Neither is there any restriction on the weight of packaged bread provided the actual weight agrees with the stated weight on the package.

Besides the main lines of bread produced at wholesale level there are small amounts of pita breads and frozen dough produced by specialty wholesale bakeries. Associated with bread production is the manufacture of wheat in New Zealand into items termed as buns and rolls. Historically buns referred to sweet raisin or currant buns. However, with the advent of hamburger buns after World War II and their present dominance in this part of the market, the terminology now has variable interpretations. Relative to bread, those products account for approximately 8% by value of the market (New Zealand Dept. of Statistics 1991). What perhaps is more significant is that the value of that market grew, from $NZ17.3 million to $NZ25.9 million in the four years from 1987 to 1991. (New Zealand Dept. of Statistics 1991).

Retail Bakeries in New Zealand comprise two distinct groups:

Hot Bread Shops. These are usually independently owned, stand-alone bakeries which bake bread, buns and rolls and some pastries and sweet goods.

Instore Bakeries. These are to be found within supermarkets and produce a similar range of products.

Bread types produced cover the whole range from French type country breads, different types of pan breads through to hearth breads of almost endless variety. Once again, multigrain breads are common. There is no reliable information available as to the relative market shares of hot bread shops and instore bakeries and wholesale bakeries. However some estimates (New Zealand Association of Bakers, Personal Communication) put the wholesale bakeries at above 80%, with the instore bakeries having a larger proportion of the remainder than hot bread shops. Instore bakeries and hot bread shops use various combinations of scratch, premix and frozen dough to produce their products. There are no reliable statistics as to the proportions of these which are used. The total bread market in New Zealand including buns and rolls in 1991 was valued at $NZ326 million at retail level (New Zealand Dept. of Statistics 1991).

Pastry Products are produced by both wholesale and retail bakeries. A full range of products is produced, with wholesale bakeries tending to specialize in frozen products (both baked and unbaked).

In common with Australians, New Zealanders eat a substantial number of meat pies. In 1991 the value of this market was reported as $NZ36.2 million or 23% of the total cake, bun and pastry market (excluding takeaways) (New Zealand Dept. of Statistics 1991). Production is characterized by a large number of small bench type operations with only 3 or 4 automated plants at strategic locations. Crust formulations vary but commonly contain 50% fat in the top rather flaky crust. The bottom crust usually has less fat (approx. 40%).

Pizzas. Following the trend of the markets in other developed countries, the consumption of pizzas has increased rapidly in recent years. Combined with quiches this sector of the market increased from $NZ11.3 million in 1988 to $NZ23.4 million in 1991 (NZ Dept. of Statistics 1991). A distinguishing feature of New Zealand pizzas compared with North America is that they are formed as a plain flat disc without the traditional lipped edge.

Biscuits. The total retail value of the biscuit (cookies and crackers) market in NZ was $NZ166.2 million in 1991. Of this, crackers contributed $NZ35.7 million or 21.5% (New Zealand Dept. of Statistics 1991). This makes this category similar to the total cake, bun and pastry market. Biscuit formulae and production methods tend to follow European trends rather than North American. It is only recently that soft cookie type biscuits have found a place in the sweet biscuit market traditionally dominated by crisp varieties. Wire cutting is the most common method of production.

Flour/Butter Products. By combining flour with the very high quality butter available in New Zealand, some companies are developing export markets for products using this combination. In particular these include frozen unbaked pastry, croissants, and cookie mixes. The pastry is usually in base form, though there is an increasing tendency for made-up products (unbaked) to be produced for markets in Asia, Australia and other Pacific rim countries.

Pasta. It has been reported (DSIR Crop Profiles 1992) that pasta has the fastest growth of any food product in New Zealand with a five fold increase in consumption in the four years to 1991 (Table 11-8). However this was from a very low base and even in 1991 the consumption was estimated to only be 1 kg per head per annum. The market indications are that this increase is continuing, fueled by the introduction by local manufacturers of fresh pasta and a wide range of imported

Table 11-8. New Zealand Pasta Consumption

Year	Domestic	Imported	Total
1986	976	2,306	3,282
1987	1,095	2,568	3,663
1988	1,405	3,124	4,529
1991	4,896	13,811	18,707

Source: New Zealand Department of Statistics

products. There is only one major producer of dried pasta in New Zealand. This company is located in the South Island and mills its own flour from locally grown durum wheat. The pasta itself is produced from a relatively dry dough (30% water) using a single screw extruder followed by controlled drying.

Grains for Specialty Breads. The increasing usage of kibbled grain in bread has resulted in attention being given to the characteristics required for such use. These are, a distinctive appearance (e.g., purple wheat) and the rate and quantity of water uptake. Purple wheat is referred to earlier in the text. For grains to be successfully incorporated into bread doughs, they need to be fully hydrated before the dough is processed. Grains which are not hydrated create unacceptable mouth feel in the final product. They also attract moisture from the surrounding crumb, thus accelerating the staling process. The quantity of water taken up by the grains also affects the yield of final product. For those reasons, grains used for specialty breads need to absorb as much water as possible as rapidly as possible. The latter is particularly important when short time doughs such as the CBP are used. The above characteristics receive particular attention from New Zealand bakers.

Breakfast Cereals. Wheat is by far the most preferred grain in New Zealand for ready to eat breakfast cereal, being more than three times preferred to any other grain (New Zealand Dept. of Statistics 1991).

Stockfeed. During the years of industry regulation, the feed market acted as a demand reservoir for undergrade milling wheat. In the free market which now exists, the amount of undergrade wheat has been reduced considerably and stockfeed mills are more inclined to contract for wheat to be grown which meets their particular needs or buy "free" (uncontracted) wheat depending on the relative cost of maize and barley with particular reference to energy values. Most stockfeed in New Zealand is produced for the poultry and horse breeding industries as well as for the dairy industry on a seasonal basis for calf rearing. Feed lots for beef are almost nonexistent in New Zealand.

The Outlook

The emphasis being placed by nutritionists on the desirability of cereals as a major portion of balanced diets is likely to mean the consumption of wheat-based products will be maintained, if not increased in New Zealand. There is some evidence in the market that food companies which specialize in cereals are adopting marketing strategies which take more advantage of the above emphasis. With the support of national guidelines promoted by the government, such strategies should result in long term benefits to the wheat growing and utilizing industry. The insignificance, by world standards, of wheat growing in New Zealand means that attention to customer requirements and the consequent specialization, offer the best chance of success. Evidence for this is the recent announcement of the export of purple wheat to Australia.

References

Australian Handbook of Breadbaking. 1989. Breadmaking processes and operations. Chapter 5. p33-65. Published by TAFE Educational Books in association with Bread Research Institute of Australia Inc.

Crofts, H.J. 1989. On defining a winter wheat. Euphytica. 44:225-234.

Crop & Food Research Institute Digest, Spring, 1992.

Crosbie, G., Miskelly, D.M., Dewan, T. 1990. Wheat quality for the Japanese flour milling and noodle industries. J.Western Australia Department of Agriculture. 3:83-88.

Department of Scientific and Industrial Research, 1992, DSIR Crop Profiles, New Zealand.

Huang, S.D., Miskelly, D.M. 1991. Steamed bread - a popular food in China. Food Australia. 43:346-351.

Macindoe, S.L., and Walkden Brown, C. 1968. In "Wheat breeding and varieties in Australia". Science Bulletin No.76 New South Wales Department of Agriculture p1.

Miskelly, D.M. 1988. Noodle and soft wheat quality for South East Asia. In: Proceedings of the thirty-eighth conference of the Royal Australian Chemical Institute - Cereal Chemistry Division. 91-95.

New Zealand Department of Statistics Agricultural Census. 1990.

New Zealand Department of Statistics. 1991. Household Expenditure & Income Survey.

New Zealand Wheat Review. 1979. No. 14.

Quail, K.J., McMaster, G.J., Wootton, M. 1991a. Flat bread production. Food Aust. 43:155-57.

Quail, K.J., McMaster, G.J., Wootton, M. 1991b. Flour quality tests for selected wheat cultivars and their relationship to Arabic bread quality. J. Sci. Fd. Agric. 54:99-110.

Index